T0225721

Model Predictive Control for AC Motors

Yaofei Han · Chao Gong · Jinqiu Gao
Editors

Model Predictive Control for AC Motors

Robustness and Accuracy Improvement Techniques

 Springer

Editors
Yaofei Han
National Maglev Transportation
Engineering R&D Center
Tongji University
Shanghai, China

Chao Gong ⓘ
School of Automation
Northwestern Polytechnical University
Xi'an, Shaanxi, China

Jinqiu Gao
School of Automation
Central South University
Changsha, Hunan, China

ISBN 978-981-16-8068-7   ISBN 978-981-16-8066-3   (eBook)
https://doi.org/10.1007/978-981-16-8066-3

This Springer imprint is published by the registered company Springer Nature Singapore Pte Ltd.
The registered company address is: 152 Beach Road, #21-01/04 Gateway East, Singapore 189721,
Singapore

Preface by Yaofei Han

This book organizes and integrates the results of the three main authors Dr. Yaofei Han, Chao Gong and Jinqiu Gao in the field of model predictive control (MPC) used in AC motors. Importantly, the achievements that the authors have made are introduced. The first chapter of this book is for readers who are new in this area. Chapters 2–5 show professional results concerning the specific problems of MPC when they are used in motor control. Hence, the readers must have at least basic knowledge of MPC and AC motors.

Considering that the previous work was finished independently, the authors of this book have tried their best to arrange the contents so as to make them as coherent as possible. But it should be noticed that in order to integrate the relevant results, some special editing operations have been employed. To make the readers understand this book simply, the following points need to be addressed:

(a) The modeling of the particular AC motors only appears once when they are needed for the first time. In the rest sections, they will be directly referred to in case of redundancy.

(b) The symbols and abbreviations in one chapter only apply to the corresponding chapter because they are defined as needed. It is normal that the description of one variable in one chapter is different from that in another.

MPC has been becoming increasingly popular in the field of AC motor control, but there are still many problems influencing the control performance. The authors will continuously focus on improving the robustness and accuracy of the MPC methods in the future. Besides, the authors welcome any discussion, suggestions and criticisms of the readers.

Shanghai, China
October 2021

Yaofei Han

Preface by Chao Gong

Model predictive control (MPC) is gaining more and more attention in the field of AC motor control at present. However, the robustness and accuracy of the MPC controllers are inclined to be influenced in the motor drive applications because of the complicated working environment and conditions. Consequently, it is valuable and significant to investigate how to enhance the robustness and accuracy of the MPC controllers used in the AC motors.

The most commonly used AC motors in engineering include induction motors (IM) and permanent magnet synchronous motors (PMSM). Besides, the wound field synchronous motors (WFSM) are witnessing a rising trend due to the novel excitation structures. This book treats these three kinds of motors as the research objects. More than one MPC strategies are focused on in this book, which include finite control set MPC (FCS-MPC), continuous control set MPC (CCS-MPC), model predictive current control (MPCC), model predictive speed control (MPSC) and multi-objective MPC.

In order to improve the robustness of the MPC controllers, disturbance observers are designed in Chap. 2. By compensating for the disturbances calculated by the observers, the impacts of the internal disturbances can be avoided completely. In Chap. 3, the real-time inductances of the AC motor are identified by using parameter observers. By using the real-time parameter for MPC, the robustness of the controller against parameter variations increases notably. Overall, this book studies two directions to solve the disturbance issue, that is, disturbance observer-based and parameter identification-based strategies. As for the topic of the accuracy of the MPC controllers, a numerical solution-based predicting plant model is developed innovatively for the surface-mounted PMSM. And, a sub-step PPM is proposed for the general PMSMs. The proposed PPMs are able to ensure high prediction accuracy of the MPC controllers in low-control-frequency situations. Moreover, flux linkage observation, delay compensation and linearization methods are adopted to improve the prediction accuracy of MPC. Overall, the methods to improve the robustness and accuracy of MPC controllers are comprehensively introduced, which can be regarded as the main contribution of this book.

This book clearly introduces the up-to-date MPC methods developed by the authors in recent years. Regarding the errors and inadequacies in the thesis, I sincerely request readers and peers to criticize and correct them.

Xi'an, China   Chao Gong
October 2021

Contents

About the Editors

Dr. Yaofei Han was born in Henan, China. He received the M.S. in 2005 and his Ph.D. in 2010, in power electronics and drives from China University of Mining and Technology respectively. He had been an associate professor at Henan University of Urban Construction since 2012, served in this capacity from 2010 to 2019. He is currently an associate professor of power electronics and electrical drives at National Maglev Transportation Engineering R&D Center, Tongji University, Shanghai, China. His research interests include multi-level power converter for power conversion and motor control, high-efficiency converter for renewable power conversion system.

Dr. Chao Gong was born in February 1991. He obtained his bachelor, master, and Ph.D. degrees in 2014, 2016, and 2021, respectively. He is going to be a tenure-track professor with the School of Automation, Northwestern Polytechnical University. His research interests include electric vehicle powertrains, motor design and control.

Ms. Jinqiu Gao was born in Shaanxi province in P.R. China, on January 7, 1996. She received the bachelor and master degrees in electrical engineering from Northwestern Polytechnical University, Xi'an, China, in 2017 and 2020, respectively. She is currently working toward the Ph.D. degree in control science and engineering with the Central South University, Changsha, China. Her research interests include fault diagnosis for traction motor, power electronics and motion control.

Chapter 1
Model Predictive Control for AC Motors

Yaofei Han, Chao Gong⊙, and Jinqiu Gao

This chapter briefly introduces the basic knowledge of model predictive control (MPC) at first. Then, the AC motors focused on in this book are presented. Thirdly, the MPC controllers used in AC motor drives are summarized, and the common problems that still need to be studied are discussed. These lay the foundation for the rest chapters. Finally, the implementations of one typical MPC method in MATLAB, which is used in permanent magnet synchronous motor (PMSM) drives, is presented, fully exerting the guiding function of this chapter.

1.1 Basic Knowledge of MPC

1.1.1 History of MPC

MPC is an advanced control method to control the process when certain constraints are met in process control. Model predictive controllers rely on dynamic models of the process, the most common being linear empirical models obtained through system identification. MPC was originally developed for chemical applications to control the instantaneous changes of dynamic systems with hundreds of inputs and outputs, And subject to constraints [1]. Since the 1980s, it has been used in the processing industry of chemical plants and oil refineries. Therefore, MPC is an effective control method that emerged and developed in the course of industrial practice. Moreover, it

Y. Han
National Maglev Transportation Engineering R&D Center, Tongji University, Shanghai 201804, China

C. Gong (✉) · J. Gao
School of Automation, Northwestern Polytechnical University, Xi'an 710072, China

has also been used in power system balance models and power electronics for over thirty years.

The development of modern control concepts can be traced back to Kalman's work in the early 1960s, when he tried to determine when a linear control system can be said to be optimal [2]. The most classic optimized controller is the linear quadratic regulator (LQR). LQR is designed to minimize the unconstrained quadratic objective function of states and inputs. But it has little influence on the development of control technology in the process industry, because its formula and the nonlinearity of the actual system have no limitations. In the late 1970s, various applications of MPC were reported in the process industry. Although the ideas of Receding Horizon Control (RHC) and MPC can be traced back to the 1960s, after the first paper on Generalized Predictive Control (GPC) was published in the 1980s [3], people began to be interested in this field. Among them, dynamic matrix control (DMC) has been designed to solve the typical multi-variable limited control problems in the petroleum and chemical industries, while the purpose of GPC is to provide a new adaptive control alternative. The basic themes of these two theories are the use of dynamic models of processes to predict the effects of future control actions, which are determined by minimizing prediction errors under operational constraints. At each sampling moment, the optimization is repeated with the latest information from the process. These algorithms are both algorithms and heuristics, and take advantage of the development of digital computers.

The original DMC algorithms represent the first-generation MPC technology, and later second-generation MPCs such as Quadratic Dynamic Matrix Control (QDMC) appeared. Although the QDMC algorithm provides a systematic way to incorporate input and output constraints, there is no clear way to deal with infeasible solutions. To solve this problem, engineers from Shell, Adersa, and Setpoint have developed a new version of the MPC algorithm, namely, identification and command (IDCOM) [4]. In the late 1980s, engineers from the French Shell Research Institute developed the Shell multivariable optimizing controller (SMOC). It is described as a bridge between the state space and the MPC algorithm. IDCOM and SMOC represent the third generation of MPC technology. In 1998, single multivariable control architecture (SMCA) and DMC technology were combined to create DMC-plus, generating the fourth generation of MPC technology. The characteristics of this generation of technology include: windows-based graphical user interface, multiple optimization levels to solve the user interface, multiple optimization levels to solve the priority control target, and improved recognition technology based on the prediction error method [5].

1.1.2 Implementations of MPC

MPC is a form of control in which the current control action is calculated by solving a finite horizon open-loop optimal control problem online at each sampling instant, using the current state of the plant as the initial state. It uses an existing model, the current state of the system and the future control variables to predict the future output of the system, and then compares it with our expected system output to get a loss used for evaluation.

Assuming that there is a state space equation with high accuracy that can describe the system, it is obvious that the future state can be predicted based on the future input of the current state. Therefore, a loss function containing two items can be defined: (1) the distance between the predicted future state and the expected future state and (2) the modulus length of the input change. Obviously, we want to optimize the loss function for future inputs so that these two items are the smallest, that is, the predicted state is closest to the expected state, and the input will not change too drastically. Different from PID control, it can be clearly seen that the process of MPC calculation input is essentially an optimization problem. Since it is an optimization problem, constraints can be added, which is not possible with proportional integral (PI) controller. Therefore, the characteristic of MPC is that it is optimized for the current time block each time, and then optimized for the time block at the next time, so that future events can be predicted and control measures can be taken accordingly. MPC is a kind of optimization control problem dedicated to a longer time span, or even infinite time, and decomposed into a number of shorter time span or finite time span optimization control problems, and to a certain extent, it still pursues the optimal control problem.

The realization process of MPC has three key steps, which are generally called three basic principles, namely prediction model, rolling optimization and feedback correction. The basic principles of them are introduced as follows:

(1)   Predictive model: Predictive model is the basis of MPC. Its main function is to predict the future output of the system based on the historical information and future input of the object. There is no strict restriction on the form of the prediction model. Traditional models such as equation of state and transfer function can all be used as prediction models. For linear stable systems, non-parametric models such as step response and impulse response can also be used directly as predictive models.

(2)   Rolling optimization: MPC determines the control effect through the optimization of a certain performance index. This performance index also relates to the future behavior of the process, which is determined by the future control strategy based on the predictive model. But the optimization is not done offline once, but repeatedly online. This is the meaning of rolling optimization and the fundamental difference between MAC and traditional optimal control. Therefore, predictive control does not use an optimized performance index that is the same as the global one, but at each moment there is a local optimized performance index relative to that moment.

(3)   Feedback correction: In order to prevent model mismatch or environmental interference from deviating from the ideal state of control, at the new sampling moment, the actual output of the object is first detected, and this real-time information is used to correct the model-based prediction results, and then proceed. New optimization. Due to the feedback correction process applied to

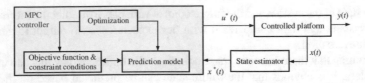

Fig. 1.1 Principle block diagram of MPC

the model, the predictive control has a strong ability to resist disturbance and overcome the uncertainty of the system. Predictive control is not only based on models, but also uses feedback information. Therefore, predictive control is a closed-loop optimization control algorithm [6].

During the control process, there is always a desired reference trajectory, and the time is the current time. The controller combines the current measured value and the prediction model to predict the output of the system in the future time domain. By solving the optimization problem that satisfies the objective function and various constraints, a series of control sequences in the control time domain (starting from the vertical axis of the coordinate system) are obtained, and the first element of the control sequence is regarded as the actual control quantity of the controlled object. When it comes to the next moment, repeat the above process, and complete the constrained optimization problems in such a rolling manner to achieve continuous control of the controlled object.

The principle block diagram of MPC is shown in Fig. 1.1, including three modules: MPC controller, controlled platform and state estimator. The MPC controller combines the predictive model, objective function and constraint conditions to optimize the solution, and obtains the optimal control sequence $u^*(t)$ at the current moment, which is input to the controlled platform and generate output $y(t)$, and the controlled platform controls according to the current control quantity, and then input the current state quantity observation value $x(t)$ to the state estimator. The state estimator estimates the state quantities that cannot be obtained by sensor observation. The commonly used methods are Kalman filter, particle filter, and so on. The estimated state quantity $x^*(t)$ is input to the MPC controller, and the optimization solution is performed again to obtain the control sequence for a period of time in the future. This cycle constitutes a complete model predictive control process.

1.1.3 Understanding MPC in View of Control

MPC belongs to the intersection of optimization and control, and can be understood from the perspective of optimal control. Therefore, MPC can be regarded as an optimization method to solve the control problem, or to give instructions to the controller by solving the optimization problem.

Optimal control emphasizes "optimality". In order to ensure optimality, optimal control generally needs to be optimized in the entire time domain. Several types of solutions commonly used in optimal control are: (1) variational method (2) maximum value principle (3) dynamic programming. variational method and maximum value principle generally can only deal with linear models and cannot contain more complicated constraints, while dynamic programming is actually a relatively advanced exhaustive method in solving optimal control problems, and its calculations are complicated. The degree is often very high. Due to the excessive emphasis on optimality, optimal control exposes two problems: (1) It is difficult to solve nonlinear and complex constraints; (2) It is necessary to accurately understand the system model. MPC retreats and only considers the next few time steps, sacrificing optimality to a certain extent. In reality, the actions of the controller often have stronger real-time requirements, and the control involving optimization problems is often more time-consuming, which is a major shortcoming of the current MPC.

1.1.4 Applications of MPC

MPC is commonly used in the complex, constrained control systems. Large-scale systems, fast systems, low-cost systems and nonlinear systems are new trends in the development of MPC research and application fields. In recent years, in many fields such as advanced manufacturing, energy and aerospace, etc., MPC has been adopted to solve constraint optimization control problems (e.g., supply chain management for semiconductor production, and material manufacturing, building energy-saving control, urban sewage treatment, flight control, satellite attitude control, etc.) The following lists several main applications of MPC:

(1) Chemical industry

The success of MPC in complex chemical processes in the past thirty years has fully demonstrated its huge potential for handling complex constrained optimization control problems in the industry.

(2) Air conditioning system

It has become a trend that the MPC algorithm is used to improve the cooling effect of the air-conditioning system and help develop more energy-efficient and efficient air-conditioning.

(3) Automotive industry

Due to the high technical threshold and the relatively limited computing power and storage resources of the chips used by mass-produced controllers due to cost considerations, the application of MPC in the automotive field is still in its early stages. In 2018, General Motors (GM) successfully applied MPC technology to GM's mass production controllers for engine torque control and gearbox ratio control, and achieved good fuel-saving effects.

(4)   Power electronics

MPC algorithms are very popular in the field of power electronics, such as motor control and converter control. It helps increase the dynamics of these systems while decreasing the design complexity.

1.2   MPC for AC Motors

1.2.1   Introduction of AC IMs and SMs

The most popular AC motors used in engineering include IMs and synchronous motors (SM). The biggest difference between an IM and an SM is whether their rotor speed is consistent with the stator rotating magnetic field. For the SM, the rotor speed is the same as the stator rotating magnetic field. Otherwise, it is called IM. Besides, the stator windings of the SMs and IMs are the same, and the difference lies in the rotor structure of the motor. The rotor of the IM is a short-circuit winding, including squirrel cage-type and winding-type IMs (see Fig. 1.2), which generates current by electromagnetic induction. The rotor of the SMs is with windings that can be supplied with direct current to form a constant magnetic field (wound field SM) or PMSMs, which are depicted in Fig. 1.3. IMs are the most widely used motors because they have the advantages of high reliability and low cost. At present, about 90% of electric-powered machinery are IMs in engineering. However, the shortcomings of the IMs are also very obvious: low efficiency and low power factor. In comparison, the advantages of the SMs include high efficiency, high power density and good stability, but the cost of the SMs is relatively higher.

In Fig. 1.3, the field windings of the wound field SM (WFSM) are fed by slip rings and brushes, which is the traditional form of a WFSM. However, the mechanical abrasion problem will reduce the service life of the system. To solve the problem, many up-to-date excitation techniques that abandon the slip rings and brushes have

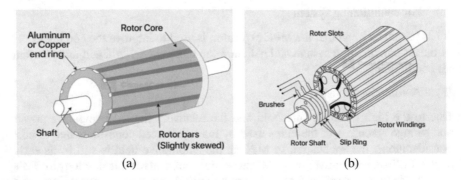

(a)   (b)

Fig. 1.2 Rotor structures of IMs. **a** Squirrel cage-type IM. **b** Winding-type IM [Google Image]

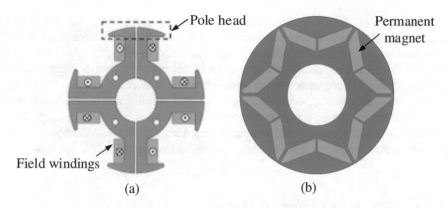

Fig. 1.3 Rotor structures of SMs. **a** WFSM. **B** PMSM

been developed, and the motors that adopt the new excitation techniques can be regarded as the novel WFSMs. Specifically, (1) a rotary transformer whose secondary side is connected with the rotating rectifier is employed to replace the excitation generators in [7]. For this technique, the primary side of the transformer is motionless and receives power from the external power source, while the secondary side rotates together with the machine rotor. (2) Considering that the rotary transformers can add significant shaft length to the machine, more compact WFSM structures based on harmonic excitation principle have been developed. In [8], the third harmonics used for excitation are generated by the separate harmonic windings embedded in the stator slots. (3) In addition to the inductive coupling approaches, a wireless power transfer excitation system based on capacitive coupling principle has arisen recently [9], of which machine structure is shown in Fig. 1.4. This scheme can offer larger output electromagnetic torque and less-distorted air-gap magnetic field.

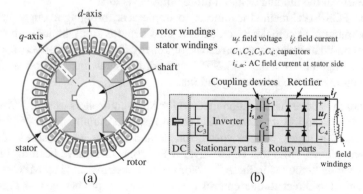

Fig. 1.4 WFSM based on capacitive coupling. **a** Machine structure. **b** Equivalent circuits of excitation components

Regardless of the IMs and the SMs (including the novel SMs), the mathematical models of them have been established and studied since they were given birth. One reason why the MPC controller is suitable for controlling the AC motors is that the mathematical models of the IMs and SMs are very clear and intuitive at present. Without losing extra time to establish the models of the AC motors and by directly using the existing models to design the MPC controllers, the controller design process becomes simple. It is even not necessary for the engineers to grasp professional knowledge of motor and automatic control theories. Hence, AC motor control provides a good application background for the development of MPC theories.

1.2.2 MPC Methods for AC Motors

There are two implementation methods of MPC in the AC motor control operations. One extends on conventional field-oriented control (FOC) by replacing the control loops with an MPC-based controller but still retaining the modulator. For this method, because the modulator can output any voltage vectors by using the sine pulse width modulation (SPWM) or space vector pulse width modulation (SVPWM) techniques, it is usually called continuous control set model predictive control (CCS-MPC). The other eliminates modulators altogether, with the outputs of MPC-based controller directly selecting the optimal inverter switch states from the finite candidates control voltage vectors, which is well-known as finite control set model predictive control (FCS-MPC). As shown in [10], the comparative features of the CCS-MPC and the FCS-MPC can be summarized as follows:

(1)   The torque and current ripples of the CCS-MPC method are notably smaller than those of the FCS-MPC methods. This represents that the CCS-MPC method shows better stable performance characteristics, which benefits from the use of the modulator and infinite voltage vectors.
(2)   The FCS-MPC method is simpler to implement, bringing about remarkable computational complexity reduction. That benefits from the look-up table of the candidate voltages, which can be obtained offline. Comparatively speaking, the computations of the CCS-MPC methods are much more complicated when executing the modulation algorithms.

Apart from the above classification method for MPC controllers used in AC motors, it needs to be mentioned that the MPC controllers are initially developed to regulate the variable states, such as speed, torque and current, etc., instead of the PI controllers. Hence, from this perspective, MPC controllers have been designed to achieve various functions in the AC motor control systems. In [11], MPC is incorporated into the direct torque control (DTC) methods, predicting and regulating the electromagnetic torque and flux of the PMSMs. This method is well known as model predictive torque control (MPTC). Paper [12] proposes a model predictive power control (MPPC) method for a permanent magnet synchronous generator to enhance the steady-state performance of the system. In addition, model predictive speed controllers (MPSC) are incorporated into the machine control topologies to achieve high-performance speed control dynamics in [13]. Another popular application is model predictive current control (MPCC), in which the currents are set as the targeting control objectives (TCO) [14]. In comparison with MPSC, the current ripples of the MPCC become lower without markedly sacrificing the dynamic performance. As far as those MPC methods are concerned, they can be grouped into two groups: single-objective method (MPPC, MPSC and MPCC) and multi-objective method (MPTC). The differences between them are that for the single-objective method, there is only one kind of targeting variable evaluated by the cost function, while there are at least two kinds of variables that need to be evaluated by the cost function for the multi-objective method.

1.2.3 Common Problems

Since there have been many scholars studying the MPC methods used in the AC motor drives for over thirty years, the relevant technologies must become mature. However, the topic is still hot, indicating that there must be some crucial issues to be solved. At present, the research concerning MPC methods used in the AC motor drives mainly focuses on the following issues: robustness and anti-disturbance capability, prediction accuracy improvement, weighting factor tuning, calculation burden reduction and theory improvement.

(a) Robustness and anti-disturbance capability

A common problem for the MPC methods used in the AC motor drives is that the system parameter variation issue influences the control performance. Unluckily, parameter variations are common for AC motors because their working conditions are usually complicated. This places great demands on the robustness of the MPC controller. In order to improve robustness of predictive control against parameter mismatch, two different approaches have been developed. Firstly, some researchers regard the predictive errors caused by the mismatched parameters as disturbances, and then compensate them by using disturbance observers [15]. Secondly, the online parameter identification techniques are incorporated into the MPC to tackle the parameter mismatch problem directly [12]. Comparatively, the disturbance observers are able to detect not only the general disturbances caused by parameter mismatch but also the system nonlinearities and even external disturbances, so it is attracting more attentions at present. For example, [15] proposes a disturbance observer based on feedback compensation to solve the inductance and bus voltage variation problem. Wang et al. [16] uses sliding mode observers (SMO) that have the advantages of robustness against disturbances, low sensitivity to the system parameter variations and fast response to solve the parameter mismatch and uncertainty problems.

In addition to the disturbances caused by the internal parameter variations, external factors, such as hardware ageing, high-frequency noises and load variations, etc., can also degrade the control performance of the MPC methods. One main reason for this phenomenon is that the bandwidth of a MPC-controller-based system is usually high so as to let the high-frequency disturbances pass through the system. Comparatively speaking, the PI-controller-based systems are more inclined to avoid external disturbances because the bandwidth of them can be regulated by adopting proper proportional and integral factors. To enhance the anti-disturbance capacity of the MPC controllers, an effective solution is to observe the disturbances and then compensate them [15]. Besides, if the external disturbances can be eliminated completely, the control performance of the MPC controllers will increase as well. In this aspect, Dr. Lu has proposed to use sensor-compensation methods to reject the disturbances caused by the ageing sensors [17].

(b)   Prediction accuracy improvement

The core of an MPC method is prediction. Prediction accuracy is the basis of high control performance. There are several factors influencing the accuracy of the prediction process, which include the machine model, nonlinearity of the motor, control period and delay, etc. Firstly, although the mathematical models for the AC motors are ready-made for dozens of years, they just represent the general properties of the motors. Usually, the concrete properties such as iron saturation and hysteresis loss are assumed to be ignored when establishing the models used for MPC control. Undoubtedly, to increase the modeling accuracy contributes to improving the prediction accuracy, and this has been addressed in [18]. Secondly, MPC can be best implemented for the systems that accept a representation by a linear model with constraints, because, in that case, most of the optimization process can be moved offline [19], leading to the fact that the standard MPC design methods usually require a linear plant model. But for the AC motors, there is a lack of efficient linearization approaches, degrading the prediction accuracy to some extent (detailed in Chap. 5). Thirdly, most of the MPC approaches utilize the forward Euler discretization algorithm to discretize the model of the AC motors, which is implemented based on one/several control periods. As for this discretization strategy, the motor states are assumed to change in a linear trend within one/several control periods. However, considering that the motor states change nonlinearly in practice, when the control period is short, the prediction accuracy is high, but if the control period is long, the prediction accuracy will degrade significantly (detailed in Chap. 4). Finally, the MPC algorithms are usually executed in the digital signal processors (e.g., DSP TMS320F28335). For a typical digital control process, in which the motor states used for prediction are measured in the present control period but the calculated switching states are applied to the machine at the beginning of the next period, there is one-sampling time delay inevitably [20]. Even for the unconventional digital control process of which sampling, calculation and actuation operations are finished within one control period, it always takes time between the sampling and the actuation. In this case, calculation delay exists. To improve the prediction accuracy, the above issues need to be solved effectively.

(c)   Weighting factor tuning

As illustrated in Sect. 1.2.2, MPC methods contain multi-objective strategies, such as MPTC control. As for the multi-objective MPC methods, at least two kinds of variables with completely different properties (torque and flux for the MPTC) need to be evaluated in the cost function. However, the combination of different variables in a single cost function is not a straightforward task, and the weighting factors in relation to each control target should be employed to settle the differences and tune the importance between those variables. Unfortunately, there are few analytical or numerical methods to design the parameters, and they are determined following the empirical approaches such as [11], which places great demand on the designers.

(d)   Calculation burden reduction

As for the FCS-MPC strategy, the candidate voltages need to be substituted into the predicting plant model to predict the future states. For the two-level inverters, there are only seven candidate voltages, which generates a relatively small amount of calculations. But in terms of the multilevel inverters, there are usually dozens of candidate voltages. The prediction process not only consumes much time but also takes up many hardware resources. In addition to the above case, the calculation burden for the multi-step-prediction-based FCS-MPC methods is heavy as well. Nowadays, to simplify the FCS-MPC methods used in the multilevel inverters-based drives and multi-step predictions process is a hot topic in the area of Ac motor control. For example, [21] presents two reduced-complexity FCS-MPC methods to reduce the calculation burden for the multilevel inverters.

(e)   Theory improvement

The problem of analyzing the stability of one optimization-based controller is always a difficult problem. Unlike the traditional controllers such as the PI or sliding mode controllers, the automatic control theories can be directly employed to analyze or design the controller/system. For instance, because the transfer function of a MPC-based AC motor drive or the Lyapunov function of the MPC controller cannot be obtained, new theories must be developed to analyze the stability of the controller/system. Simon and Löfberg [22] proposes to use the Karuhn-Kuhn-Tucker (KKT) conditions to give a sufficient stability condition of the MPC controller for particular linear systems. However, there are still few stability analysis techniques used for the nonlinear systems such as the AC motor drives. Hence, there is still a long way to go before mature theories are developed.

1.3   Implementations in MATLAB of Typical MPC

Finite control set model predictive current control (FCS-MPCC) is popular in engineering. As for the FCS-MPCC method, one MPC controller is used to replace two PI current regulators in the traditional topology, which is clearly explained in [23]. This part takes a two-level based PMSM as an example, the codes of an implementation method of a typical FCS-MPCC controller are given. It deserves to be mentioned that the codes in this part are independent from the research results in the later chapters.

Instead, it will exert the function of providing guidelines for the new researchers who are new to the field.

```
%****************************************************************
                %FCS-MPCC Codes for PMSM in MATLAB 2018a
%****************************************************************
                           Function Definition
function [Sa, Sb, Sc] = fcn(id, iq, idref, iqref, wm, theta)
%Sa, Sb, Sc are switching state; id, iq are measured currents;
idref, iqref are reference
%currents; wm is speed, theta is rotor position.
%****************************************************************
                          %Parameter Definition
Ld=0.1; %d-axis inductance
Lq=0.1; %q-axis inductance
Rs=0.1; %stator resistance
T=1.0e-4; %control period
f=0.1; %magnetic flux
p=3; %pole pairs
Udc=100; %DC-bus voltage
A=[(Ld-T*R)/Ld,wm*T*Lq*p/Ld;
    -wm*T*Ld*p/Lq,(Lq-T*R)/Lq];
B=[T/Ld,0;0,T/Lq];
C=[0;-wm*T*f*p/Lq];
%****************************************************************
                        %Voltage Vector Definition
v1=[0;0;0];
v2=[-Udc/3;- Udc /3;2* Udc /3];
v3=[- Udc /3; Udc *2/3;- Udc /3];
v4=[-2* Udc /3; Udc /3; Udc /3];
v5=[2* Udc /3;- Udc /3;- Udc /3];
v6=[ Udc /3;-2* Udc /3; Udc /3];
v7=[ Udc /3; Udc /3;-2* Udc /3];
v8=[ Udc /3; Udc /3 Udc /3];
v=[v1,v2,v3,v4,v5,v6,v7,v8];
%****************************************************************
                        %Switching State Definition
states = [0 0 0;0 0 1;0 1 0;0 1 1;1 0 0;1 0 1;1 1 0;1 1 1];
%****************************************************************
                      %Current Transformation Matrix
Park=[cos(theta),sin(theta);
    -sin(theta),cos(theta);];
Clarke=(2/3)*[1,-0.5,-0.5;
        0,sqrt(3)/2,-sqrt(3)/2];
%****************************************************************
                      %Prediction and Optimization
idqk=[id;iq];%measurement
gn=zeros(1,8);
uu=zeros(2,1);
for i=1:1:8
    u=Park*(Clarke*v(:,i));
    uu(1,1)=u(1,1);
    uu(2,1)=u(2,1);
    idqkk=A*idqk+B*uu+C;%prediction
```

```
        gn(i)=(iqref-idqkk(2,1))^2+(idref-idqkk(1,1))^2;%cost func-
tion calculation
end
[~, x_opt] = min(gn);%The best switching state selection
%****************************************************************
                                %Switching State Output
Sa = states(x_opt,1);
Sb = states(x_opt,2);
Sc = states(x_opt,3);
```

1.4 Summary

This chapter introduces the basic knowledge of MPC and the significance of this research. The main contributions can be summarized as follows:

(1) The history, implementations and applications of MPC are reviewed briefly.
(2) The structures of IMs, PMSMs and WFSMs are introduced, which are the control objects of this book.
(3) The common problems of MPC used in the AC motor drives are presented, illustrating the significance of the research results in the later chapters.
(4) The implementations of a typical FCCS-MPCC method used in PMSM in MATLAB are given, providing guidelines for the new researchers.

References

1. S.J. Qin, T.A. Badgwell, A survey of industrial model predictive. Control. Eng. Pract. **93**(316), 733–764 (2003)
2. R. Kalman, "A new approach to linear filtering and prediction problems," in *Trans. ASME, J. Basic Engineering*, pp. 35–45, 1960.
3. D.W. Clarke, C. Mohtadi, P.S. Tuffs, Generalized predictive control—part i: the basic algorithm. Automatica **23**, 137–148 (1987)
4. P. Grosdidier, B. Froisy, M. Hammann, "The IDCOM-M controller," in *Proceedings of the 1988 IFAC workshop on model based process control*, (Oxford, 1988)
5. Model predictive control: history and development," Int. J. Eng. Trends Technol. **4**(6), 2600–2602 (Jun 2013)
6. X. Xu, Z. Mao, Analysis and research on predictive control based on hammerstein model. Control Theory Appl. **17**(4), 529–532 (2000)
7. J. Tang, Y. Liu, N. Sharma, Modeling and experimental verification of high-frequency inductive brushless exciter for electrically excited synchronous machines. IEEE Trans. Industry Appl. **55**(5), 4613–4623 (2019)
8. G. Jawad, Q. Ali, T. A. Lipo, B. Kwon, "Novel brushless wound rotor synchronous machine with zero-sequence third-harmonic field excitation." IEEE Trans. Magnetics **52**(7), 1–4 (July 2016), Art no. 8106104.
9. D.C. Ludois, J.K. Reed, K. Hanson, Capacitive power transfer for rotor field current in synchronous machines. IEEE Trans. Power Electron. **27**(11), 4638–4645 (2012)

10. A.A. Ahmed, B.K. Koh, Y. Lee, Continuous control set-model predictive control for torque control of induction motors in a wide speed range. Electric Power Components and Syst. **46**(19), 2142–5158 (2018)

11. M. Preindl, S. Bolognani, Model predictive direct torque control with finite control set for pmsm drive systems, part 1: maximum torque per ampere operation. IEEE Trans. Industr. Inf. **9**(4), 1912–1921 (2013)

12. S. Kwak, U. Moon, J. Park, Predictive-control-based direct power control with an adaptive parameter identification technique for improved afe performance. IEEE Trans. Power Electron. **29**(11), 6178–6187 (2014)

13. E. Fuentes, C.A. Silva, R.M. Kennel, MPC implementation of a quasi-time-optimal speed control for a PMSM drive, with inner modulated-FS-MPC torque control. IEEE Trans. Industr. Electron. **63**(6), 3897–3905 (2016)

14. M. Yang, X. Lang, J. Long, D. Xu, Flux immunity robust predictive current control with incremental model and extended state observer for PMSM drive. IEEE Trans. Power Electron. **32**(12), 9267–9279 (2017)

15. S. Kang, J. Soh, R. Kim, K. Lee, S. Kim, Robust predictive current control for IPMSM without rotor flux information based on a discrete-time disturbance observer. IET Electr. Power Appl. **13**(12), 2079–2089 (2019)

16. B. Wang, Z. Dong, Y. Yu, G. Wang, D. Xu, Static-errorless deadbeat predictive current control using second-order sliding-mode disturbance observer for induction machine drives. IEEE Trans. Power Electron. **33**(3), 2395–2403 (2018)

17. J. Lu, Y. Hu, J. Liu, Z. Wang, All current sensor survivable IPMSM drive with reconfigurable inverter. IEEE Trans. Industr. Electron. **67**(8), 6331–6341 (2020)

18. Š. Janouš, T. Košan, J. Talla, Z. Peroutka, "Improved accuracy of model predictive control of Induction motor drive using FPGA," *2019 IEEE International Symposium on Predictive Control of Electrical Drives and Power Electronics (PRECEDE)*, 2019, pp. 1–6

19. A. Linder, R. Kennel, "Model predictive control for electrical drives," *2005 IEEE 36th Power Electronics Specialists Conference*, (Recife, 2005), pp. 1793–1799

20. C. Choi, W. Lee, Analysis and compensation of time delay effects in hardware-in-the-loop simulation for automotive PMSM drive system. IEEE Trans. Industr. Electron. **59**(9), 3403–3410 (2012)

21. I. Harbi, M. Abdelrahem, M. Ahmed, R. Kennel, Reduced-complexity model predictive control with online parameter assessment for a grid-connected single-phase multilevel inverter. Sustainability **12**(19), 7997–8020 (2020)

22. D. Simon, J. Löfberg, "Stability analysis of model predictive controllers using Mixed Integer Linear Programming," *2016 IEEE 55th Conference on Decision and Control (CDC)*, 2016, pp. 7270–7275

23. J. Gao, C. Gong, W. Li, J. Liu, Novel compensation strategy for calculation delay of finite control set model predictive current control in PMSM. IEEE Trans. Industr. Electron. **67**(7), 5816–5819 (2020)

Chapter 2
Observer-Based Robustness Improvement for FCS-MPCC Used in IMs

Yaofei Han, Chao Gong, and Jinqiu Gao

This Chapter proposes a sliding mode (SM) disturbance observer based finite control set model predictive current control (FCS-MPCC) strategy to improve the control performance of induction motors. FCS-MPCC method is achieved based on the machine model, leading to the fact that the parameters have great impacts on the control performance, especially the steady-state characteristics. In this Chapter, the predicting model for induction motors (IMs) is established first. Simultaneously, a current-type flux observer is constructed according to the machine properties. Then, based on the predicting plant model (PPM), the disturbances caused by parameter mismatch are analyzed, posing the necessity of developing effective measures to compensate the disturbances. In this Chapter, a novel SM disturbance observer based on hyperbolic switching function is established to detect the real-time disturbances arising from parameter mismatch. Meanwhile, a Lyapunov function is constructed to analyze the stability of the observer and it is proven that there exist the internal gain factors to ensure the SM observer to be stable. Finally, the proposed disturbance observer is incorporated into the FCS-MPCC implementation process to eliminate the influence of parameter mismatch. The simulation is conducted on a three-phase IM to verify the performance of the proposed SM disturbance-observer-based FCS-MPCC strategy.

Y. Han
National Maglev Transportation Engineering R&D Center, Tongji University, Shanghai 201804, China

C. Gong (✉) · J. Gao
School of Automation, Northwestern Polytechnical University, Xi'an 710072, China

© The Author(s) 2022
Y. Han et al. (eds.), *Model Predictive Control for AC Motors*,
https://doi.org/10.1007/978-981-16-8066-3_2

2.1 Problem Descriptions

As illustrated in Chap. 1, FCS-MPC strategies have been widely used in the IM drives. Usually, the prediction process of an FCS-MPC controller is based on the machine model, leading to the fact that the system control performance is closely related to the machine's electrical parameters, including the winding resistance and inductance, etc. In practice, the machine parameter values for FCS-MPC are usually provided by the suppliers on the nameplate. However, the parameters vary greatly during operations due to the ambient conditions (e.g. temperature and magnetic field). On this ground there must exist errors between the offline parameters and the real-time ones, leading to current divergence and static errors [1]. In order to improve the robustness of predictive control against parameter mismatch, two different approaches have been developed. Firstly, some researchers regard the predictive errors caused by the mismatched parameters as disturbances, and then compensate them by using perturbation observers [2–4]. Secondly, the online parameter identification techniques are incorporated into the MPC to tackle the parameter mismatch problem [5–7]. Comparatively, the disturbance observers are able to detect not only the general disturbances caused by parameter mismatch but also the system nonlinearities and even external disturbances, so it is attracting more attentions at present. Paper [8] proposes a disturbance observer based on feedback compensation to solve the inductance and bus voltage variation problem. [9, 10] use sliding mode observers (SMO) that have the advantages of robustness against disturbances, low sensitivity to the system parameter variations and fast response to solve the parameter mismatch and uncertainty problems, laying the ground for the SMO based disturbance computation and compensation techniques. However, it needs to be mentioned that although the second order SMO with low chattering has been successfully applied in [9], the algorithm is complicated to implement in practice. Besides, the SMO is particularly designed for the deadbeat MPC, and there are few researches concerning the sliding mode disturbance observers used for the FCS-MPC applications. As far as the observer in [10] is concerned, it is only proposed for the mechanical load torque rather than electrical parameter estimation for the permanent magnet synchronous motors (PMSMs). Thus, the sliding mode electrical observers that are characterized by easier implementation are highly required for the FCS-MPC based IMs.

This Chapter proposes a sliding mode perturbation observer–based FCS model predictive current control (FCS-MPCC) for IMs to improve the robustness against the machine resistance and inductance mismatch phenomenon. It needs to be mentioned that the disturbance observer that uses a new constructed-function (CF) for chattering attenuation is innovatively used for IM FCS-MPCC situations. After establishing the mathematical model of the machine, the impacts of parameter mismatch on the control performance (static errors) of an FCS-MPCC controller is analyzed in detail. Then, a SMO based on CF is designed to obtain the system disturbance. Compared to the high order SMO in [9], the new observer is easier to achieve without using the differential-based reaching law. In order to ensure the new SMO to remain stable, a Lyapunov function is constructed and the stability conditions are calculated. Finally, by integrating the estimated disturbance into the prediction model, the robust FCS-MPCC algorithm is achieved. Compared to the traditional FCS-MPCC algorithms, the current divergence and static errors will get suppressed in the resistance and inductance.

2.2 Implementation of FCS-MPCC and Impacts of Parameter Mismatch on Control Performance

A discrete PPM is the prerequisite for achieving an IM FCS-MPCC algorithm. In this section, the mathematical model of an IM is established firstly. To obtain the rotor flux used for feedback regulation, a novel numerical solution based current-type flux observer is presented secondly. Then, the implementation procedures are given in detail. Finally, the relationship between the control performance and parameter errors is theoretically discussed, posing the necessity to employ a disturbance observer for compensation.

2.2.1 State-Space Model of IM

State-space model is explicit and easy to describe the general behaviors of a multi-variable system. In order to estimate the future current states, the electrical dynamics of IMs should be employed for analysis, and the differential equations in MT-axis frame are as follows, where the iron saturation and eddy current are assumed to be negligible.

$$\frac{di_{sM}}{dt} = -\frac{R_s}{L_s}i_{sM} + \omega_e i_{sT} - \frac{L_m}{L_s}\frac{di_{rM}}{dt} + \frac{L_m}{L_s}\omega_e i_{rT} + \frac{u_{sM}}{L_s} \tag{2.1}$$

$$\frac{di_{sT}}{dt} = -\omega_e i_{sM} - \frac{R_s}{L_s}i_{sT} - \frac{L_m}{L_s}\omega_e i_{rM} - \frac{L_m}{L_s}\frac{di_{rT}}{dt} + \frac{u_{sT}}{L_s} \tag{2.2}$$

$$\frac{di_{rM}}{dt} = -\frac{L_m}{L_r}\frac{di_{sM}}{dt} + \frac{L_m}{L_r}\Delta\omega i_{sT} - \frac{R_r}{L_r}i_{rM} + \Delta\omega i_{rT} \tag{2.3}$$

$$\frac{di_{rT}}{dt} = -\frac{\Delta\omega L_m}{L_r}i_{sM} - \frac{L_m}{L_r}\frac{di_{sT}}{dt} - \Delta\omega i_{rM} - \frac{R_r}{L_r}i_{rT} \tag{2.4}$$

$$T_e = pL_m(i_{sT}i_{rM} - i_{sM}i_{rT}) = p\frac{L_m}{L_r}\psi_r i_{sT} \tag{2.5}$$

where i_{sM} and i_{sT} are stator currents. u_{sM} and u_{sT} are stator voltage. ω_e and $\Delta\omega$ are the synchronous speed and slip speed, respectively. i_{rM} and i_{rT} are the rotor currents.

R_s and L_s represent the stator resistance and inductance, respectively. R_r and L_r are the rotor resistance and inductance. L_m is the mutual inductance between the stator and rotor windings. T_e is the electromagnetic torque. ψ_r is the rotor flux and p is the number of pole pairs. Substitute (2.3) and (2.4) into (2.1) and (2.2), respectively and then, discretize them in a time step of T (sampling time) to calculate the future states as follows:

$$i_{sM}^{k+1} = (1 - C_1 L_r R_s) i_{sM}^k + C_1 [(L_s L_r \omega_e^k - L_m^2 \Delta \omega^k) i_{sT}^k$$
$$+ (\omega_e^k - \Delta \omega^k) L_r L_m i_{rT}^k + L_m R_r i_{rM}^k + L_r u_{sM}^k] \tag{2.6}$$

$$i_{sT}^{k+1} = (1 - C_1 L_r R_s) i_{sT}^k + C_1 [(L_m^2 \Delta \omega^k - L_s L_r \omega_e^k) i_{sM}^k$$
$$- (\omega_e^k - \Delta \omega^k) L_r L_m i_{rM}^k + L_m R_r i_{rT}^k + L_r u_{sT}^k] \tag{2.7}$$

where the superscripts $k + 1$ and k represent the future and current states, respectively. C_1 is a constant, and it satisfies that $C_1 = \frac{T}{L_s L_r - L_m^2}$. In the process of prediction, L_m can be set to zero by ignoring the mutual coupling [11].

Differing from synchronous machines, the flux rotating speed in the IM is not in sync with the mechanical speed, leading to the fact that the measured speed and position using external sensors cannot be directly used for coordinate transformation and prediction. Simultaneously, because the rotor flux is induced by the magnetic field of the stator, the amplitude of the rotor flux should be regulated in an FCS-MPCC strategy. On these grounds it is crucial to accurately obtain the amplitude and phase angle of the rotor flux and the synchronous speed for IM control. Now, there exist two main flux observation methods, that is, current-type observer [12] and voltage-type observer [13]. Comparatively, the latter approach accuracy is low over the low-speed range due to the internal integrators. Therefore, a current-type observer is adopted in this Chapter. As in [12], according to (2.1)–(2.5) and the flux descriptions, the rotor flux can be derived as:

$$\psi_r = \frac{L_m i_{sM}}{T_r s + 1} \tag{2.8}$$

where T_r is the electrical time constant of the rotor windings, and it equals $\frac{L_r}{R_r}$. And the slip speed is:

$$\Delta \omega = \frac{L_m i_{sT}}{T_r \psi_r + 1} \rightarrow \omega_e = p \omega_r + \Delta \omega \tag{2.9}$$

where ω_r represents the rotor speed. It can be noted that once the rotor flux is only related to the M-axis current, and once ψ_r is calculated, the slip speed can be obtained subsequently. Further, the phase angle θ of the rotor flux, which is used for coordinate transformation, can be calculated by:

$$\theta = \int \omega_e dt = \int (p\omega_r + \Delta\omega)dt \tag{2.10}$$

It can be seen from (2.8) that the relationship between the rotor flux and M-axis current can be treated as a first-order inertia element. In this chapter, in order to calculate ψ_r, the method based on solving the differential equation is adopted, namely,

$$\psi_r(t) = L_m i_{sM}(1 - e^{-\frac{t}{T_r}}) \tag{2.11}$$

In the control period between t_k and t_{k+1}, where $0 \le t \le T$, i_{sM}^k can be directly substituted into (2.11) to calculate ψ_r at any moment. Considering that the rotor flux shifts constantly, we can treat the typical value at $t_k + 0.5\,T$ as the output of the flux observer within the current cycle, that is,

$$\psi_r^k = L_m i_{sM}^k (1 - e^{-\frac{T}{2T_r}}) \tag{2.12}$$

When an FCS-MPCC algorithm is implemented, the candidate control voltages applied to the stator windings should be directly substituted into the predicting model one by one to predict the next-step states. In terms of a two-level inverter, a total of seven phase voltage vectors that are denoted as are among the alternatives:

$$\mathbf{v}_{s_a s_b s_c} = \begin{bmatrix} u_a \\ u_b \\ u_c \end{bmatrix} = \frac{U_{dc}}{3} \begin{bmatrix} 2 & -1 & -1 \\ -1 & 2 & -1 \\ -1 & -1 & 2 \end{bmatrix} \begin{bmatrix} s_a \\ s_b \\ s_c \end{bmatrix} \tag{2.13}$$

where $[s_a, s_b, s_c]^T$ is the switching states. They are $[0, 0, 0]^T, [1, 0, 0]^T, [1, 1, 0]^T, [0, 1, 0]^T, [0, 1, 1]^T, [0, 0, 1]^T$ and $[1, 0, 1]^T$, which can also be denoted as $v_1, \ldots, v_7$. U_{dc} is the DC-bus source voltage. $[u_a, u_b, u_c]^T$ are the terminal phase voltages. By the use of transformation, the control voltage sets used for prediction can be expressed as:

$$\begin{bmatrix} u_{sM}^k \\ u_{sT}^k \end{bmatrix} = \sqrt{\frac{2}{3}} \begin{bmatrix} \cos\theta & \dfrac{\sqrt{3}\sin\theta - \cos\theta}{2} & \dfrac{-\sqrt{3}\sin\theta - \cos\theta}{2} \\ -\sin\theta & \dfrac{\sin\theta + \sqrt{3}\cos\theta}{2} & \dfrac{\sin\theta - \sqrt{3}\cos\theta}{2} \end{bmatrix} \cdot \mathbf{v}_{s_a s_b s_c} \tag{2.14}$$

The block diagram of the FCS-MPCC implementation is shown in Fig. 2.1. At the kth period, the FCS-MPCC implementation procedures are:

(a) Measurement: the phase currents and speed are measured by using current and position sensors.

(b) *abc*/*MT* transformation: the measured phase currents are transformed to the *MT*-axis currents according to the rotor position estimated in the last period.

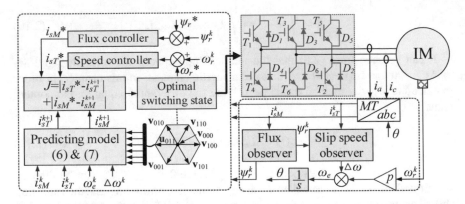

Fig. 2.1 Block diagram of the FCS-MPCC implementation

(c) Observation: use the immediate currents to calculate the rotor flux, slip speed, synchronous speed and position.

(d) Prediction: use the immediate speed and currents to estimate the future current states i_{sM}^{k+1} and i_{sT}^{k+1} for all the seven candidate voltage vectors.

(e) Evaluation: substitute the seven the predicted values into the cost function (2.15) and determine the optimal voltage vector and the corresponding switching states.

$$J = \left| i_{sM}^* - i_{sM}^{k+1} \right| + \left| i_{sT}^* - i_{sT}^{k+1} \right| \tag{2.15}$$

where i_{sT}^* and i_{sM}^* are the reference currents.

(f) Actuation: apply the optimum switching state to the drive system.

2.2.2 Impacts of Parameter Mismatch on Performance

Paper [1] analyses the sources (inductance and flux disturbance, etc.) of the current static errors for PMSM drives, laying the ground for theoretically analyzing IMs, though it neglects the impact of the winding resistance. This part will comprehensively explain the impacts of resistance, mutual and self-inductance mismatch on the FCS-MPC static errors.

The measured resistance and inductance used for FCS-MPCC controller are denoted as R_{s0}, L_{s0}, L_{r0} and L_{m0}. Taking the M-axis current as an example, the estimated value at t_k is denoted as i_{sM0}^{k+1}, and the predicting model can be written as:

$$i_{sM0}^{k+1} = (1 - C_{1\_0}L_{r0}R_{s0})i_{sM}^k + C_{1\_0}[(L_{s0}L_{r0}\omega_e^k - L_{m0}^2\Delta\omega^k)i_{sT}^k$$
$$+ (\omega_e^k - \Delta\omega^k)L_{r0}L_{m0}i_{rT}^k + L_{m0}R_{r0}i_{rM}^k + L_{r0}u_{sM}^k] \tag{2.16}$$

where C_{1_0} relies on the offline inductance values. Further, the estimation error $\triangle i_{sM}$ can be deduced by subtracting (2.16) from (2.6) as follows:

$$\begin{cases} \triangle i_{sM} = i_{sM}^{k+1} - i_{sM0}^{k+1} = \triangle k_1 i_{sM}^k + \triangle k_2 i_{sT}^k + \triangle k_3 i_{rT}^k + \triangle k_4 i_{rM}^k + \triangle k_5 i_{rM}^k \\ \triangle k_1 = C_{1\_0} L_{r0} R_{s0} - C_1 L_r R_s \\ \triangle k_2 = C_{1\_0}(L_s L_r \omega_e^k - L_m^2 \triangle \omega^k) - C_{1\_0}(L_{s0} L_{r0} \omega_e^k - L_{m0}^2 \triangle \omega^k) \\ \triangle k_3 = C_{1\_0}(\omega_e^k - \triangle \omega^k) L_r L_m - C_{1\_0}(\omega_e^k - \triangle \omega^k) L_{r0} L_{m0} \\ \triangle k_4 = C_{1\_0} L_m R_r - C_{1\_0} L_{m0} R_{r0} \\ \triangle k_5 = C_{1\_0} L_r - C_{1\_0} L_{r0} \end{cases}$$

$$(2.17)$$

According to (2.17), it can be noted that the static error is proportional to the control period T. Usually, the switching period is small (e.g., 0.1 ms), so the current static error will be pretty small unless the parameter mismatch phenomenon is severe. Obviously, the higher the magnitude of $\triangle k_n$ ($n = 1, 2, 3, 4, 5$) is, the larger the current static error becomes. This represents that parameter mismatch will greatly reduce the steady-state estimation accuracy of an FCS-MPCC controller. Unluckily, as for the traditional FCS-MPCC algorithms ($L_{m0} = 0$), there always exists disturbance caused by parameter mismatch. Moreover, it can be seen that the estimation errors are highly related to the machine speed, so the control performance might decrease visibly in the high-speed range.

Overall, if the perturbation caused by parameter mismatch is not compensated, the static current errors will inevitably appear, lowering the control performance of a predictive controller [5]. Even worse, the system will become unstable when the parameter deviations are very large. Therefore, it is necessary to introduce a perturbation observer to detect the real-time disturbance caused by the parameter mismatch and compensate the FCS-MPCC algorithms.

2.3   Proposed Sliding Mode Disturbance Observer

The conventional sliding mode observers are achieved by the use the signum function, which is shaped like Fig. 2.2a. However, due to the inertia of a real system, its response speed is much slower than the expected property (steep trend) near the sliding mode surface (SMS). Hence, chattering arises inevitably. The reason why the SMO in [9] is complicated is that a high order sliding mode surface is adopted to suppress

Fig. 2.2 Properties of switching functions. **a** Signum function; **b** Hyperbolic function

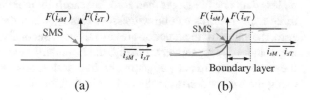

(a)   (b)

the chattering effects. In this section, a simpler SMO based on hyperbolic function shaped like Fig. 2.2b is proposed. It deserves to be mentioned that the switching function is pretty new, and because it is continuous and differentiable (modest trend) near the SMS, the chattering phenomenon can be attenuated.

2.3.1 Sliding Mode Disturbance Observer

Denote the M, T-axis disturbances caused by parameter mismatch as f_M and f_T, respectively. In virtue of the measured parameters, the IM model containing the disturbances can be rewritten as:

$$\frac{di_{sM}}{dt} = -\frac{R_{s0}}{L_{s0}}i_{sM} + \omega_e i_{sT} + \frac{u_{sM}}{L_{s0}} + f_M \tag{2.18}$$

$$\frac{di_{sT}}{dt} = -\omega_e i_{sM} - \frac{R_{s0}}{L_{s0}}i_{sT} + \frac{u_{sT}}{L_{s0}} + f_T \tag{2.19}$$

According to the sliding mode theory, the disturbance observer can be constructed as:

$$\frac{di_{sM}^*}{dt} = -\frac{R_{s0}}{L_{s0}}i_{sM}^* + \omega_e i_{sT} + \frac{u_{sM}}{L_{s0}} + k_M F(\overline{i_{sM}}) \tag{2.20}$$

$$\frac{di_{sT}^*}{dt} = -\omega_e i_{sM} - \frac{R_{s0}}{L_{s0}}i_{sT}^* + \frac{u_{sT}}{L_{s0}} + k_T F(\overline{i_{sT}}) \tag{2.21}$$

where i_{sM}^* and i_{sT}^* are estimated stator currents. $\overline{i_{sM}}$ and $\overline{i_{sT}}$ represent the errors between the estimated and real currents, and $\overline{i_{sM}} = i_{sM}^* - i_{sM}$, $\overline{i_{sT}} = i_{sT}^* - i_{sT}$. k_M and k_T are the gain factors. $F(\overline{i_{sM}})$ and $F(\overline{i_{sT}})$ are the hyperbolic function, that is,

$$\begin{bmatrix} F(\overline{i_{sM}}) \\ F(\overline{i_{sT}}) \end{bmatrix} = \begin{bmatrix} \dfrac{e^{m \cdot \overline{i_{sM}}} - e^{-m \cdot \overline{i_{sM}}}}{e^{m \cdot \overline{i_{sM}}} + e^{-m \cdot \overline{i_{sM}}}} \\ \dfrac{e^{m \cdot \overline{i_{sT}}} - e^{-m \cdot \overline{i_{sT}}}}{e^{m \cdot \overline{i_{sT}}} + e^{-m \cdot \overline{i_{sT}}}} \end{bmatrix} \tag{2.22}$$

where m is a constant for boundary layer (as in Fig. 2.2b) regulation, and it should be less than one but larger than zero. By carefully looking at the switching function, three main features can be witnessed. Firstly, when the estimated errors approach zero, the values of the function are zero as well. Secondly, the upper and lower output are 1 and -1, respectively. Finally, the boundary layer needs to be in accord with the system properties by adjusting m. In practice, this parameter can be obtained by using the trial-and-error method. In this chapter, $m = 0.78$. If the system reaches a

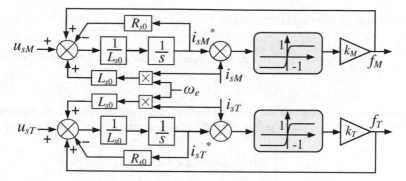

Fig. 2.3 Block diagram of proposed SMO

stable state, the actual disturbances can be obtained by the following equations:

$$\begin{bmatrix} f_M \\ f_T \end{bmatrix} = \begin{bmatrix} k_M F(\overline{i_{sM}}) \\ k_T F(\overline{i_{sT}}) \end{bmatrix}$$

(2.23)

The block diagram of the proposed observer is shown in Fig. 2.3.

2.3.2 Stability Analysis

In order to construct Lyapunov function for stability analysis, the sliding mode surfaces of M, T-axis currents are denoted as S_M and S_T, respectively. They should satisfy the following conditions:

$$\mathbf{S} = \begin{bmatrix} S_M \\ S_T \end{bmatrix} = \begin{bmatrix} \overline{i_{sM}} \\ \overline{i_{sT}} \end{bmatrix} = 0$$

(2.24)

The Lyapunov function V is as follows:

$$V = \frac{1}{2}\mathbf{S} \cdot \mathbf{S}^T = \frac{1}{2}\overline{i_{sM}}^2 + \frac{1}{2}\overline{i_{sT}}^2$$

(2.25)

No doubt that the value of V is larger than zero. Only by proving that the derivative of V is less than zero can we conclude that the proposed observer is stable. Take the time derivative of (2.25) and it can be deduced that:

$$\frac{dV}{dt} = \overline{i_{sM}} \cdot \frac{d\overline{i_{sM}}}{dt} + \overline{i_{sT}} \cdot \frac{d\overline{i_{sT}}}{dt}$$

(2.26)

Substitute (2.18)–(2.21) into (2.26), and then we can obtain that:

$$\frac{dV}{dt} = -\frac{R_{s0}}{L_{s0}}\overline{i_{sM}}^2 - \frac{R_{s0}}{L_{s0}}\overline{i_{sT}}^2 + (k_M F(\overline{i_{sM}}) - f_M)\overline{i_{sM}} + (k_T F(\overline{i_{sT}}) - f_T)\overline{i_{sT}}$$
(2.27)

Obviously, the values of the first and second terms are less than zero. Therefore, in order to make the system stable, the following equations need to be satisfied:

$$\begin{cases} (k_M F(\overline{i_{sM}}) - f_M)\overline{i_{sM}} < 0 \\ (k_T F(\overline{i_{sT}}) - f_T)\overline{i_{sT}} < 0 \end{cases}$$
(2.28)

According to the signs (positive or negative) of the M, T-axis current errors, (2.28) can be further derived as:

$$\begin{cases} k_M < \dfrac{f_M}{F(\overline{i_{sM}})}, & if\,\overline{i_{sM}} > 0 \\ k_M < -\dfrac{f_M}{F(\overline{i_{sM}})}, & if\,\overline{i_{sM}} < 0 \end{cases} \rightarrow k_M < -\left|\dfrac{f_M}{F(\overline{i_{sM}})}\right|$$
(2.29)

$$\begin{cases} k_T < \dfrac{f_T}{F(\overline{i_{sT}})}, & if\,\overline{i_{sT}} > 0 \\ k_T < -\dfrac{f_T}{F(\overline{i_{sT}})}, & if\,\overline{i_{sT}} < 0 \end{cases} \rightarrow k_T < -\left|\dfrac{f_T}{F(\overline{i_{sT}})}\right|$$
(2.30)

Combing (2.29) and (2.30), the observer stability condition can be summarized as:

$$k_M, k_T < \min\left(-\left|\frac{f_M}{F(\overline{i_{sM}})}\right|, -\left|\frac{f_T}{F(\overline{i_{sT}})}\right|\right)$$
(2.31)

Because $F(\overline{i_{sM}})$ and $F(\overline{i_{sT}})$ are between -1 and 1, theoretically, even by making the magnitudes of k_M and k_T extremely large, the observer cannot always keep stable, especially when $\overline{i_{sM}}$ and $\overline{i_{sT}}$ are near zero. But in the real applications, the current estimation errors will fluctuate mostly around the sliding surfaces with small variations with the range of tolerance [14]. Define the smallest tolerance as η,

$$\eta = \min(|\overline{i_{sM}}|, |\overline{i_{sT}}|)$$
(2.32)

Then, the minimum switching function value is:

$$\min F = \frac{e^{m\cdot\eta} - e^{-m\cdot\eta}}{e^{m\cdot\eta} + e^{-m\cdot\eta}}$$
(2.33)

Hence, the observer gain can be set as:

$$k_M = k_T = \min(-\left|\frac{f_M}{\min F}\right|, -\left|\frac{f_T}{\min F}\right|) \tag{2.34}$$

In theory, there exist k_M and k_T making the observer stable as long as the disturbance values are not exaggeratedly large (where the system is unstable). During the control process, although $\overline{i_{sM}}$ and $\overline{i_{sT}}$ are possible to be less than the pre-set value of η, the proposed SMO can re-converge again once their values increase. Hence, although the proposed observer is asymptotically stable, the disturbances can be estimated.

After estimating the disturbances caused by the parameter mismatch using the proposed SMO, the disturbances can be directly incorporated the FCS-MPCC PPM for compensation. By applying Euler implementation to (2.18) and (2.19), the predicting model used in practice is:

$$i_{sM}^{k+1} = \frac{L_{s0} - T R_{s0}}{L_{s0}} i_{sM}^k + T\omega_e i_{sT}^k + \frac{T u_{sM}^k}{L_{s0}} + T f_M^k \tag{2.35}$$

$$i_{sT}^{k+1} = -T\omega_e i_{sM}^k + \frac{L_{s0} - T R_{s0}}{L_{s0}} i_{sT}^k + \frac{T u_{sT}^k}{L_{s0}} + T f_T^k \tag{2.36}$$

where f_M^k and f_T^k are the disturbances at t_k observed by the SMO. The implementation (simulation) diagram of the improved FCS-MPCC method is depicted in Fig. 2.4. As to this controller, although the offline parameters are still used for future state prediction, the disturbances can be removed.

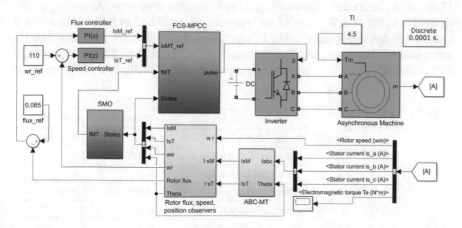

Fig. 2.4 Implementation diagram of the proposed SMO based FCS-MPCC

2.4 Verifications

Simulation is conducted on a three-phase IM whose real parameters and structure are assumed to be consistent with those in Table 2.1 to verify the effectiveness of the proposed sliding mode disturbance observer-based FCS-MPCC. Considering that the disturbances principally affect the static errors during control, the steady-state control performance is mainly discussed in this section, though the dynamics are given. In simulation, the control frequency is set to 10 kHz.

By comparing the PPM (2.35) and (2.36) with the comprehensive ones (2.6) and (2.7), without considering the disturbance terms, the rotor parameters (rotor inductance and resistance) are removed totally when the mutual effect is ignored. On this ground the rotor parameters can be treated to be one hundred percent (100%) mismatched in terms of an FCS-MPCC algorithm. As for the stator inductance and resistance, they are inclined to encounter mismatch problems during control as well. For the sake of comprehensiveness, two typical cases are analyzed in this part. *Case 1*: the measured stator inductance and resistance are 90% lower than the actual ones. *Case 2*: the measured stator inductance and resistance are 50% higher than the actual ones. For each case, different working conditions (40 rad/s, 80 rad/s and 110 rad/s under load and no-load) are taken into account.

2.4.1 Case 1

(a) *FCS-MPCC without disturbance observer*

Figure 2.5 shows the dynamic speed and torque performance characteristics of the FCS-MPCC controller without using any disturbance observers. According to the dynamics in Fig. 2.5, the simulation setup can be described as follows. Between 0 and 3 s, the machine speeds up from 0 to 40 rad/s, after which the reference speed is set as 80 rad/s. Between 6 and 12 s, the machine is expected to be controlled to remain 110 rad/s, and at 12 s, the motor decelerates to 80 rad/s. Finally, from 15 s, the speed

Table 2.1 Measured IM parameters and structure

Parameters	Value	Unit
Stator inductance L_s	0.12	H
Stator resistance R_s	0.065	Ω
Mutual inductance L_m	0.075	H
Rotor inductance L_r	0.095	H
Rotor resistance R_r	0.05	Ω
Friction coefficient F	0.003	–
Rated load T_{L_rated}	4.5	Nm
Rated speed ω_{r_rated}	110	rad/s

Fig. 2.5 Speed and torque dynamic performance of machine

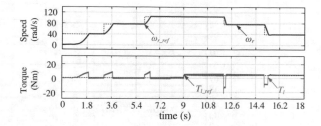

reference is set to 40 rad/s. In terms of the external load, before and after 12 s, it is set to 0 and 4.5 Nm, respectively. Before leaving Fig. 2.5, it should be noticed that firstly, the machine has fast dynamics regardless of acceleration and deceleration. Specifically, the rise time for the acceleration processes (0–40 rad/s, 40–80 rad/s and 80–110 rad/s) is nearly 1.85 s, 0.65 s and 0.45 s, respectively. The deceleration is a little faster than the acceleration process. In detail, the settling time is about 0.3 s.

Figure 2.6 depicts the steady-state performance of the FCS- MPCC at the speed of 40 rad/s under no-load (between 2.5 s and 3 s) and rated load (between 17 s and 17.5 s) conditions. In the no-load conditions, the speed can remain at 42 rad/s, so the static error rate (*SER*) is 5%. As far as the rotor flux is concerned, the *SER* is 2.4%. Comparatively, the speed and flux SER is 3% and 8.2%, respectively. This is interesting that the static error for speed decreases, but it increases for the flux after the external load is imposed on the rotor shaft. This happens mainly because that the flux is more closely related to the load (current) than the speed. Because currents are not the final control targets, we just give the current ripples of the i_{sM} and i_{sT}. Comparatively speaking, the M, T-axis current ripples (*CR*) under load (3.1 A and 1.3 A) are slightly higher than those under no load (2.4 A and 1.25 A).

The steady-state performance of the system when the machine runs at 80 rad/s is illustrated in Fig. 2.7. Compared to Fig. 2.6, the magnitude of the speed static error increases to 3.2 rad/s and 1.8 rad/s under no-load and load conditions. This

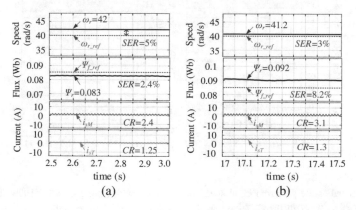

Fig. 2.6 Steady-state performance at speed of 40 rad/s with rotor parameter 100% and stator parameter 90%-lower mismatch. **a** No-load condition; **b** Rated load condition

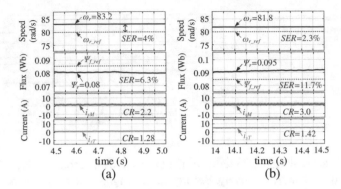

Fig. 2.7 Steady-state performance at speed of 80 rad/s with rotor parameter 100% and stator parameter 90%-lower mismatch. **a** No-load condition; **b** Rated load condition

complies with the conclusion in this chapter. In terms of the rotor flux, the *STR* witnesses a dramatical growth regardless of the load conditions. i_{sM} and i_{sT} are very similar to those in the situation that the machine speed is 40 rad/s, and they are 2.2 A and 1.28 A when the machine rotates without external load, 3.0 A and 1.42 A with rated load imposed on the shaft. Figure 2.8 demonstrates the control performance over higher speed range (11 rad/s), it can be seen that the magnitudes of the static errors experience an upward trend in comparison with the low-speed cases, while the currents nearly remain at the similar level. Overall, the traditional FCS-MPCC that ignores the mutual coupling has marked static errors over the full-speed range.

(b)   FCS-MPCC with SM disturbance observer

Figure 2.9 a, b show the estimated disturbances using the proposed SMO when the machine rotates at 40 rad/s under no load and load, respectively. At the moment,

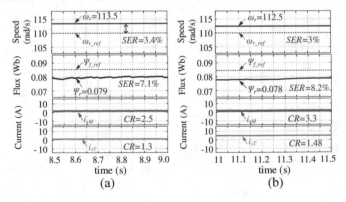

Fig. 2.8 Steady-state performance at speed of 110 rad/s with rotor parameter 100% and stator parameter 90%-lower mismatch. **a** No-load condition; **b** Rated load condition

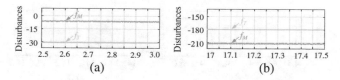

Fig. 2.9 Steady-state estimated disturbances using the proposed SMO at the speed of 40 rad/s. **a** No-load condition; **b** Rated load condition

the observed disturbances have not been incorporated into the FCS-MPCC implementations. It can be seen when the machine runs without load, the M-axis disturbance (around -5) is much smaller than the T-axis one (-31). However, in the load condition, the M-axis witnesses a sharp increase to about -180, while the T-axis disturbance is about -210. Although the estimated values are pretty large, when they are substituted into (2.35) and (2.36) for prediction, they need to multiply by the sampling period (0.001 s). Therefore, the estimated values are reasonable. Overall, these results represent that the disturbances caused by rotor and stator parameter mismatch is closely related to the load states.

After incorporating the SMO into the FCS-MPCC control process, the control performance characteristics at the speed of 40 rad/s are shown in Fig. 2.10. Compared to the results in Fig. 2.6, the speed and flux static errors have been dramatically reduced regardless of the working conditions. In detail, the SER for speed and rotor flux is 0.5% and 0.057% respectively under no load, and it is 0.75% and 0 respectively in the load conditions. These represent that the proposed observer is able to improve the system control performance. Apart from the static errors, the current ripples experience a visible decrease as well. In simulation, the M- and T-axis current ripples become 0.6 and 0.55 A under no load.

Figure 2.11a, b show the estimated disturbances at the speed of 80 rad/s and Fig. 2.11c, d depict the control performance after compensation. Obviously, the disturbances are much larger than those in Fig. 2.9, so it is further proven that

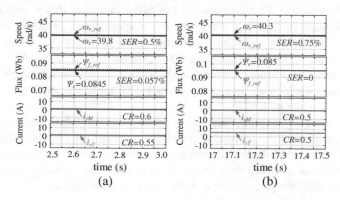

Fig. 2.10 Steady-state performance of the machine when incorporating the proposed SMO at the speed of 40 rad/s. **a** No-load condition; **b** Rated load condition

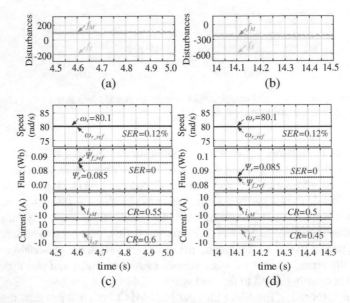

Fig. 2.11 Steady-state performance of the machine when incorporating the proposed SMO at the speed of 80 rad/s. **a** No-load-condition disturbances; **b** Rated-load-condition disturbances; **c** No-load-condition performance; **d** Rated-load-condition performance

the disturbances are highly related to the machine speed. Similar as the results in Fig. 2.10, the control performance after compensation. Obviously, the disturbances are much larger than those in Fig. 2.9, so it is further proven that the disturbances are proportionally related to the machine speed. Similar as the results in Fig. 2.10, thespeed and flux static errors are nearly zero. In terms of the M, T-axis currents, the ripples are about 0.5 A, which is truly small. Therefore, the steady-state control performance has been improved greatly when adopting the proposed observer.

In Fig. 2.12, the no-load-condition and rated-load-condition disturbances and the steady-state control performance at the speed of 110 rad/s are demonstrated. The disturbances become larger than the lower speed cases. In detail, the M-axis disturbance is about 50, while the T-axis disturbance is -780 under no load, but they become around -275 and -1180 when the external load is imposed on the rotor shaft. After the SMO is adopted for disturbance observation and compensation, the speed and flux static errors become very small (*SER* for speed and flux is 0.4% and 0 respectively under no load).

In order to clearly reveal the superiority of the proposed SMO based FCS-MPCC, Table 2.2 summarizes the speed and flux static errors and the M-axis current ripples before and after employing the disturbance observer. it deserves to be mentioned that aiming at the case in which the rotor parameters and stator parameters are 100 and 90% mismatched, the proposed FCS-MPCC has great capability.

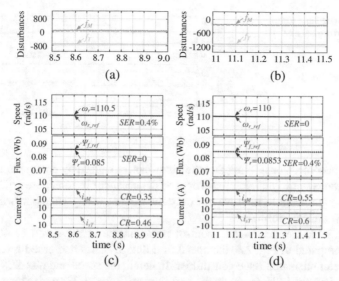

Fig. 2.12 Steady-state performance of the machine when incorporating the proposed SMO at the speed of 110 rad/s. **a** No-load-condition disturbances; **b** Rated-load-condition disturbances; **c** No-load-condition performance; **d** Rated-load-condition performance

Table 2.2 Comparison of static errors and M-axis current ripple before and after employing SMO

Speed (rad/s)	Speed static error (rad/s)		Flux static error (Wb)		Ripples of i_{sM} (A)	
	Before	After	Before	After	Before	After
40	2	0.2	0.002	0.0005	2.4	0.6
	1.2	0.3	0.007	0	3.1	0.5
80	3.2	0.1	0.005	0	2.2	0.55
	1.8	0.1	0.01	0	3.0	0.5
110	3.5	0.5	0.006	0	2.5	0.35
	2.5	0	0.007	0.0003	3.3	0.55

2.4.2   Case 2

(a)   FCS-MPCC without disturbance observer

When the stator parameters provided by the suppliers (used for control) are 50% higher than the actual ones, the dynamic speed and torque performance without using any disturbance observers are shown in Fig. 2.13. The simulation setup is the same to that in Fig. 2.5, but the dynamics are a little different. In detail, the rise time for the acceleration processes (0–40 rad/s, 40–80 rad/s and 80–110 rad/s) is 0.98 s, 0.5 s and 0.28 s, respectively, and the settling time for deceleration is even shorter. These represent that the milder the parameter mismatch phenomenon is, the better the

Fig. 2.13 Speed and torque dynamic performance of machine

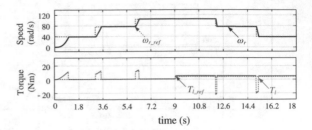

dynamics become. Considering that the disturbances caused by parameter mismatch influence the system dynamics, the proposed disturbance observer based FCS-MPCC strategy is significant to improve not only the system steady-state performance but also the dynamic performance.

In Fig. 2.14, the steady-state control performance at the speed of 40 rad/s is shown. Compared to Fig. 2.6, the speed and flux *SER* has decreased greatly when the parameter mismatch issue gets milder. In detail, the speed and flux *SER* under no load are 2.5% and 1.2% respectively, and they are 2 and 1.2% under load. In terms of the current ripples, they are 1.4 A and 1.05 A for the *M* and *T* axis respectively in the no-load condition.

Figures 2.15 and 2.16 illustrates the steady-state control performance at the speed of 80 rad/s and 110 rad/s, respectively. Similarly, as the speed increases, the magnitude of the static errors will become larger while the *SER* of them decreases. Interestingly, by contrast with Case 1, the flux fluctuations reduced markedly when the parameter mismatch phenomenon becomes wilder. At 80 rad/s, the speed and flux static errors are 1.5 rad/s and 0.002 Wb in the no-load condition, and 1.2 rad/s and 0.001 Wb, respectively. The reason why the control performance gets slightly better under load is that the impacts of disturbances caused by parameter mismatch are uncertain at different working states. In other words, At 110 rad/s, the static errors

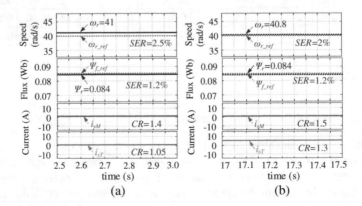

Fig. 2.14 Steady-state performance at speed of 40 rad/s with rotor parameter 100% and stator parameter 50%-higher mismatch. **a** No-load condition; **b** Rated load condition

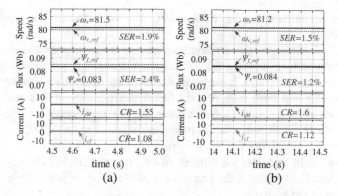

Fig. 2.15 Steady-state performance at speed of 80 rad/s with rotor parameter 100% and stator parameter 50%-higher mismatch. **a** No-load condition; **b** Rated load condition

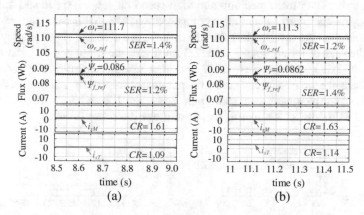

Fig. 2.16 Steady-state performance at speed of 110 rad/s with rotor parameter 100% and stator parameter 50%-higher mismatch. **a** No-load condition; **b** Rated load condition

for speed and flux are 1.7 rad/s and 0.001 Wb under no load, and they are 1.3 rad/s and 0.0012 Wb under load, respectively. As far as the M, T-axis current ripples are concerned, they are nearly 1.6 A and 1.1 A respectively regardless of the load and speed states.

Overall, the impacts of parameter mismatch are significant to both the dynamic and steady-state control performance of FCS-MPCC strategies, and meanwhile, the mismatch degree is proportionally relevant to the system performance. That is, the more severe the parameter mismatch phenomenon is, the worse the system performance becomes.

(a)   FCS-MPCC with SM disturbance observer

It needs to be mentioned that the effective disturbance compensation techniques are still needed in the less-severe parameter mismatch situations. This part shows the control performance of the FCS-MPCC method based on the proposed SM perturbation observer.

Figure 2.17 illustrates the disturbances caused by parameter mismatch before compensation and the steady-state control performance after compensation when the machine runs at 40 rad/s. It can be noticed that the disturbances are smaller than those in Fig. 2.9, and they are the M-axis disturbance is about -4 than the T-axis one is -22 under no load. In the load condition, the M-axis disturbance also increases sharply to about -170, while the T-axis disturbance is about -200. Figure 2.17c, d show that the speed and flux static errors are nearly zero despite of no-load or load situations. In addition, the current ripples become half of the values before compensation. Consequently, the proposed SMO is able to effectively deal with the disturbances in the system.

Figures 2.18 and 2.19 show control performance at the speed of 80 rad/s and 110 rad/s, respectively. Obviously, the higher the speed is, the larger the disturbances turn out to be. Over these medium- and high-speed ranges, the speed and flux static errors approaches zero as well after compensation. Meanwhile, the current ripples are smaller than the results when the SM disturbance observer is not employed.

For the sake of intuitiveness, Table 2.3 illustrates the speed and flux static errors and the M-axis current ripples before and after using the disturbance observer to compensate the disturbances caused by parameter mismatch. Aiming at the case where the rotor parameters and stator parameters are 100 and 50% mismatched, the proposed FCS-MPCC is also effective to improve the system control performance.

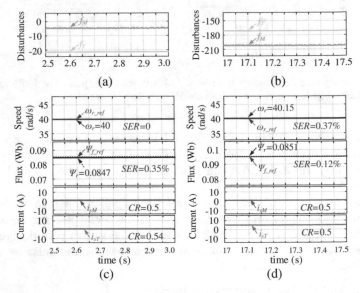

Fig. 2.17 Steady-state performance of the machine when incorporating the proposed SMO at the speed of 40 rad/s. **a** No-load-condition disturbances; **b** Rated-load-condition disturbances; **c** No-load-condition performance; **d** Rated-load-condition performance

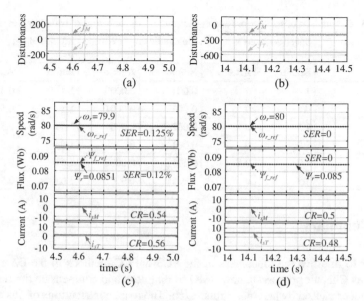

Fig. 2.18 Steady-state performance of the machine when incorporating the proposed SMO at the speed of 80 rad/s. **a** No-load-condition disturbances; **b** Rated-load-condition disturbances; **c** No-load-condition performance; **d** Rated-load-condition performance

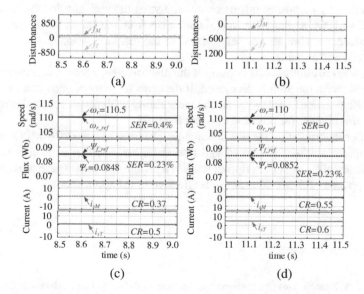

Fig. 2.19 Steady-state performance of the machine when incorporating the proposed SMO at the speed of 110 rad/s. **a** No-load-condition disturbances; **b** Rated-load-condition disturbances; **c** No-load-condition performance **d** Rated-load-condition performance

Table 2.3 Comparison of static errors and M-axis current ripple before and after employing SMO

Speed (rad/s)	Speed static error (rad/s)		Flux static error (Wb)		Ripples of i_{sM} (A)	
	Before	After	Before	After	Before	After
40	1	0	0.001	0.0003	1.4	0.5
	0.8	0.15	0.001	0.0001	1.5	0.5
80	1.5	0.1	0.002	0.0001	1.55	0.54
	1.2	0	0.001	0	1.6	0.5
110	1.7	0.5	0.001	0.0002	1.61	0.37
	1.3	0	0.0012	0.0002	1.63	0.55

2.5 Summary

In order to improve the robustness of the FCS-MPCC methods in the IM applications, this Chapter proposes a new SMO to diagnose and compensate the real-time disturbances caused by parameter mismatch. The main contributions of this chapter are as follows.

(1) After establishing the predicting model for IMs, the M, T-axis disturbances caused by parameter mismatch are analyzed theoretically for the traditional FCS-MPCC strategy. It is found that, apart from the parameter deviations, the disturbances are also closely related to the machine working states, such as speed and currents.

(2) In order to eliminate the impacts of the disturbances, a SMO observer based on hyperbolic function is developed. It deserves to be mentioned that the hyperbolic function is pretty novel, and the stability analysis procedures introduced in this chapter for this kind of SMO is new.

(3) After incorporating the SM disturbance observer into the FCS-MPCC control implementations, the speed and flux static errors are significantly reduced. Meanwhile, the current ripples witness a visible decrease. These prove that the proposed FCS-MPCC strategy is effective in IM applications.

References

1. Z. Liu, Y. Zhao, Robust perturbation observer-based finite control set model predictive current control for spmsm considering parameter mismatch. Energies **12**(19), 3711–3724 (2019)
2. J. Liu, W. Wu, H.S. Chung, F. Blaabjerg, Disturbance observer-based adaptive current control with self-learning ability to improve the grid-injected current for LCL -filtered grid-connected inverter. IEEE Access **7**, 105376–105390 (2019)
3. S. Kang, J. Soh, R. Kim, K. Lee, S. Kim, Robust predictive current control for IPMSM without rotor flux information based on a discrete-time disturbance observer. IET Electric Power Appl. **13**(12), 2079–2089 (2019)

4. X. Zhang, B. Hou, Y. Mei, Deadbeat predictive current control of permanent-magnet synchronous motors with stator current and disturbance observer. IEEE Trans. Power Electron. **32**(5), 3818–3834 (2017)
5. S. Kwak, U. Moon, J. Park, Predictive-control-based direct power control with an adaptive parameter identification technique for improved AFE performance. IEEE Trans. Power Electron. **29**(11), 6178–6187 (2014)
6. J. Holtz, J. Quan, Sensorless vector control of induction motors at very low speed using a nonlinear inverter model and parameter identification. IEEE Trans. Indus. Appl. **38**(4), 1087–1095 (2002)
7. G. Gatto, I. Marongiu, A. Serpi, Discrete-time parameter identification of a surface-mounted permanent magnet synchronous machine. IEEE Trans. Indus. Electron. **60**(11), 4869–4880 (2013)
8. M. Ali, M. Yaqoob, L. Cao, K.H. Loo, Disturbance-observer-based DC-bus voltage control for ripple mitigation and improved dynamic response in two-stage single-phase inverter system. IEEE Trans. Indus. Electron. **66**(9), 6836–6845 (2019)
9. B. Wang, Z. Dong, Y. Yu, G. Wang, D. Xu, Static-errorless deadbeat predictive current control using second-order sliding-mode disturbance observer for induction machine drives. IEEE Trans. Indus. Electron. **33**(3), 2395–2403 (2018)
10. C. Gong, Y. Hu, K. Ni, J. Liu, J. Gao, SM load torque observer-based FCS-MPDSC with single prediction horizon for high dynamics of surface-mounted PMSM. IEEE Trans. Indus. Electron. **35**(1), 20–24 (2020)
11. H. Yang, Y. Zhang, P. Huang, "Improved predictive current control of IM drives based on a sliding mode observer," *2019 IEEE International Symposium on Predictive Control of Electrical Drives and Power Electronics (PRECEDE)*, Quanzhou, China, 2019, pp. 1–6
12. W. Bu, S. Wang, C. Zu, L. Zhai, "Rotor flux estimation method of bearingless induction motor based on stator current vector orientation," 2012 IEEE International Conference on Automation and Logistics, Zhengzhou, 2012, pp. 437–441
13. M. Hinkkanen, J. Luomi, Modified integrator for voltage model flux estimation of induction motors. IEEE Trans. Indus. Electron. **50**(4), 818–820 (2003)
14. C. Gong, Y. Hu, J. Gao, Y. Wang, L. Yan, An improved delay-suppressed sliding-mode observer for sensorless vector-controlled PMSM. IEEE Trans. Indus. Electron. **67**(7), 5913–5923 (2020)

Chapter 3
Parameter-Identification-Based Robustness Improvement for FCS-MPC Used in WFSMs

Yaofei Han, Chao Gong⑩, and Jinqiu Gao

As illustrated in Chap. 1, the novel wound field synchronous motors (WFSMs) without installing slip rings and brushes are drawing increasing attention in engineering. To ensure high control performance of the novel WFSMs, this chapter proposes a robust finite control set model predictive current control (FCS-MPCC) strategy based on series of new sliding mode (SM) inductance observers. Firstly, after establishing the model of a WFSM based on capacitive coupling, the predicting plant model (PPM) is designed for FCS-MPCC control. Secondly, to eliminate the impacts of the inductance uncertainties on the FCS-MPCC control, the SM mutual inductance observer (MIO), q-axis inductance observer (QAIO) and d-axis inductance observer (DAIO) and field winding inductance observer (FWIO) are designed, which need to be implemented one by one. Then, the Lyapunov stability criterion is used to analyze the stability conditions for the observers. Moreover, considering that the observers are achieved by using the offline inductance information provided by the suppliers, the robustness against parameter mismatch is discussed at length, and an analytical method that can avoid estimation errors is developed. The proposed inductance identification techniques and FCS-MPCC method are verified by simulation which is conducted on a 580 W three-phase WFSM drive.

3.1 Problem Descriptions

When the novel WFSMs based on capacitive coupling excitation are adopted in engineering, they should have marked control performance. Because the FCS-MPCC

Y. Han
National Maglev Transportation Engineering R&D Center, Tongji University, Shanghai 201804, China

C. Gong (✉) · J. Gao
School of Automation, Northwestern Polytechnical University, Xi'an 710072, China

strategy has the advantages of fast response, simple structure and no need for tuning parameters, etc., it is well-suited for WFSM control. However, considering that the values of the machine inductances require to be directly used to construct the MPC controller, the accuracy of them will influence the control performance inevitably [1, 2]. Traditionally, the offline-tested inductances are employed to achieve the FCS-MPCC algorithms, which are usually fixed. However, the machine inductances can be influenced by the operating environment (high/low temperature and magnetic saturation), resulting in parameter variations in real applications. Hence, the traditional fixed-parameter-based FCS-MPCC strategy has weak robustness against the inductance variations. For the sake of high control performance, it is necessary to obtain the real-time values of the machine inductances. Unluckily, there are few particular studies concerning the parameter identification methods for the aforementioned unconventional capacitive-coupling-excitation-based WFSMs (see Chap. 1) controlled by the FCS-MPCC method.

Referring to the existing parameter identification methods for induction machines and PMSMs, the potential suitable ways for a WFSM can be divided into two groups: offline identification at a standstill and online identification in working conditions [3–13]. In engineering, the machine parameters are mainly detected by the offline measurement means, which can be provided by the suppliers or manufacturers [4]. These methods require to actuate the motor to tremble by injecting different presetting AC/DC current/voltage excitations under load or no-load conditions and then, the response signals are used for calculating the parameters to be tested [5]. Although the offline identification methods have shown relatively satisfactory performance in practice, they have one common disadvantage that the detected inductance and resistance cannot alter as the operating environment changes. For example, the impacts of magnetic saturation cannot be reflected in the offline measured inductance, resulting in that parameter mismatch phenomenon occurs in the saturation conditions. On this ground, the online real-time parameter detection methods are highly favored by scholars. At present, the commonly used online techniques can be grouped into five classes: model reference adaptive system (MRAS) based estimation [6], recursive least square (RLS) based estimation [7, 8], extended Kalman filter (EKF) based estimation [9], state-observer based estimation [10] and artificial intelligence (AI) based estimation [11]. Compared to the former three algorithms which are closely dependent on the system model, the latter two have much stronger robustness against the parameter variations so that the one-parameter estimation accuracy stands high even though the other parameters are mismatched. Moreover, considering that it is tedious to establish a credible database for tuning the coefficients of an AI controller, the state-observer-based identification methods become superior. For example, a Luenberger-sliding mode observer is presented in [12] to detect the armature resistance, and [13] adopts a reduced-order observer to identify the rotor time constant. However, it needs to be mentioned that most of the previous state observers have the disadvantage of low bandwidth [14], leading to the fact that the observed parameters do not comply with real-time ones precisely. In addition, few scholars focus on the robustness of the state observers against the parameter uncertainties before.

To solve the low-response-speed problem, pure SM observers that are well-known for their rapidity (high bandwidth) can be selected to detect the parameters [15]. Whereas, the method has not been widely used for multiple electrical parameter estimation processes of IMs and PMSMs, not to mention the WFSMs. Consequently, it is significant to deeply study the pure SM principle-based multi-parameter identification techniques and analytical rules (including analysis of SM observers' own robustness) for WFSMs, which can be further used to enhance the robustness of the FCS-MPCC control method.

This chapter presents a robust FCS-MPCC control method against inductance variations for the novel WFSMs based on capacitive coupling excitation, which only relies on the SM theory to precisely estimate the real-time inductances step by step. The main novelties of this part can be summarized as:

(1) Pure SM inductance observers are designed to detect the rea-time mutual and self-inductances step by step, which can be further used to achieve the robust FCS-MPCC control method. Lyapunov functions are constructed to analyze the stability of the SM observers, obtaining the explicit conditions that make the system stable.

(2) Uncertain inductances are needed to construct the proposed SM-MIO and the QAIO. Concerning this issue, the robustness of the observers against parameter uncertainties is discussed, obtaining an analytical method to accurately calculate the inductances. This has been seldom studied previously.

(3) The structure and implementations (see Fig. 3.1) of the proposed SM inductance observers-based on FCS-MPCC strategy are illustrated.

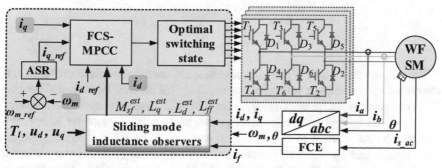

i_{d_ref}, i_{q_ref}: d, q-axis current references, i_{s_ac}: AC current of stationary part, ω_{m_ref}: speed reference, θ: position, $i_{a,b}$: armature current, FCE: field current estimation

Fig. 3.1 Block diagram of FCS-MPCC control strategy

3.2 Modeling of WFSMs

The mathematical model extracted from a physical motor not only reveals the parameters that require to be measured or estimated for control but also is the prerequisite for a model-based online parameter identification and FCS-MPCC methods. As far as the WFSMs based on capacitive coupling are concerned, the mechanical structure and equivalent circuits for the excitation components are shown in Fig. 1.4. As is depicted in Fig. 1.4b, the output of the rectifier serves as the exciting voltage u_f which is applied to the field windings directly. This means except the excitation devices, the rest parts of a novel WFSM rotor are the same as those of a traditional motor. On this ground when ignoring the excitation components, the mathematical model that demonstrates the electrical and mechanical behaviors in the direct-quadrature rotating frame for a novel WFSM can be described as [16]:

$$\frac{\mathrm{d}i_d}{\mathrm{d}t} = -\frac{R_s}{L_d}i_d + \frac{L_q}{L_d}p\omega_m i_q - \frac{M_{sf}}{L_d}\frac{\mathrm{d}i_f}{\mathrm{d}t} + \frac{1}{L_d}u_d \tag{3.1}$$

$$\frac{\mathrm{d}i_q}{\mathrm{d}t} = -\frac{L_d}{L_q}p\omega_m i_d - \frac{R_s}{L_q}i_q - \frac{M_{sf}}{L_q}p\omega_m i_f + \frac{1}{L_q}u_q \tag{3.2}$$

$$\frac{\mathrm{d}i_f}{\mathrm{d}t} = -\frac{3M_{sf}}{2L_{ff}}\frac{\mathrm{d}i_d}{\mathrm{d}t} - \frac{R_f}{L_{ff}}i_f + \frac{1}{L_{ff}}u_f \tag{3.3}$$

$$T_e = \frac{3}{2}p(L_d i_d + M_{sf}i_f - L_q i_d)i_q \tag{3.4}$$

$$\frac{\mathrm{d}\omega_m}{\mathrm{d}t} = \frac{1}{J}(T_e - B\omega_m - T_l) \tag{3.5}$$

where i_d, i_q are the stator d, q-axis currents and u_d, u_q are the d, q-axis the voltages. i_f is the excitation current, and with reference to [17–19], it needs to be estimated because it is not available for measurement. As for a WFSM based on capacitive coupling, the excitation current i_f together with u_f can be calculated based on Kirchhoff's law and the AC current of the stationary part i_{s_ac} [17]. L_d, L_q are the armature inductances, and L_{ff} is the self-inductance of field windings. The stator and rotor winding resistance are R_s and R_f, respectively. M_{sf} is the mutual inductance between the field and armature windings. ω_m is the rotor mechanical angular speed and p represents the number of pole pairs. T_e and T_l are the output electromagnetic torque and the load torque, respectively. J is the rotor inertia and B represents the damping coefficient. It is noted that considering this research mainly focuses on the impacts of the armature inductance mismatch on the FCS-MPCC performance, an appropriate assumption that T_l, J, B and R_s can be detected accurately by using offline or online strategies in practice can be made.

To achieve the FCS-MPCC algorithms, substitute (3.3) into (3.1) and apply the forward Euler discretization method to the electrical armature model of the WFSM, and the discrete PPM used for FCS-MPCC in a time step of T can be expressed as:

$$i_d(k+1) = (1 - C_1 L_{ff} R_s) i_d(k) + C_1 L_{ff} L_q p \omega_m(k) i_q(k)$$
$$+ C_1 M_{sf} R_f i_f(k) - C_1 [M_{sf} u_f(k) + L_{ff} u_d(k)] \quad (3.6)$$

$$i_q(k+1) = -\frac{T L_d p}{L_q} \omega_m(k) i_d(k) + \frac{L_q - T R_s}{L_q} i_q(k)$$
$$- \frac{T M_{sf} p}{L_q} \omega_m(k) i_f(k) + \frac{T}{L_q} u_q(k) \quad (3.7)$$

where $i_d(k)$, $i_q(k)$, $i_f(k)$ and $\omega_m(k)$ are the measured values at the kth sampling instant t_k. $i_d(k+1)$, $i_q(k+1)$, $i_f(k+1)$ and $\omega_m(k+1)$ are the estimated states. $u_d(k)$ and $u_q(k)$ represent the candidate manipulated voltages, and there are seven voltages constituting the finite set for a two-level voltage inverter. C_1 is a constant and $C_1 = \frac{2T}{2L_{ff} L_d - 3M_{sf}^2}$. Based on (3.6) and (3.7), it can be seen that there are four inductance parameters that influence the prediction accuracy of FCS-MPCC, that is, L_d, L_q, L_{ff} and M_{sf}, which need to be identified online.

3.3 SM Observer-Based Parameter Identification

As for the to-be-observed inductance parameters, theoretically, because the Eqs. (3.1), (3.2) and (3.5) contain the information of L_d, L_q and M_{sf}, each of them can be used to estimate these three parameters when using SM theory. As for the inductance of field windings, since only (3.3) contains its information, it is the only equation that is suitable for identifying L_{ff}. In this part, considering that it is possible that the QAIO might be very close to the d-axis inductance in reality, neither L_q nor L_d will be obtained using (3.5), and instead, they are observed by the use of (3.1) and (3.2), respectively. In terms of (3.5), it will be adopted for mutual-inductance estimation. On these grounds, in line with the four steps illustrated in Fig. 3.2, series of SM observers are to be developed with their stability and robustness discussed in this section.

3.3.1 Design of Sm Observers

(a) SM-MIO

Prior to establishing an SM-MIO, (3.4) needs to be substituted into (3.5) and it can be derived that:

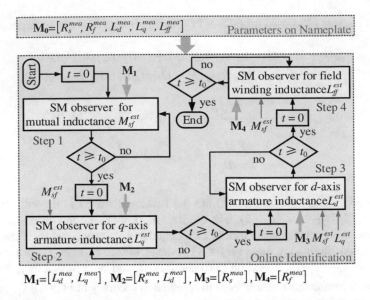

$\mathbf{M_0} = [R_s^{mea}, R_f^{mea}, L_d^{mea}, L_q^{mea}, L_{ff}^{mea}]$ Parameters on Nameplate

$\mathbf{M_1} = [L_d^{mea}, L_q^{mea}]$, $\mathbf{M_2} = [R_s^{mea}, L_d^{mea}]$, $\mathbf{M_3} = [R_s^{mea}]$, $\mathbf{M_4} = [R_f^{mea}]$

Fig. 3.2 Flow chart of proposed Multistep parameter identification strategy

$$\frac{d\omega_m}{dt} = \frac{3}{2J} p(L_d i_d + M_{sf} i_f - L_q i_d) i_q - \frac{1}{J}(B\omega_m + T_l) \tag{3.8}$$

Then, the SM observer based on (3.8) can be represented as:

$$\frac{d\omega_m^*}{dt} = \frac{3}{2J} p(L_d i_d + k_{mi} F(\overline{\omega_m}) i_f - L_q i_d) i_q - \frac{1}{J}(B\omega_m^* + T_l) \tag{3.9}$$

where ω_m^* is the estimated speed. $\overline{\omega_m}$ is the error between the estimated speed and the real speed, that is, $\overline{\omega_m} = \omega_m^* - \omega_m$. $F(\overline{\omega_m})$ represents the signum function, namely,

$$F(\overline{\omega_m}) = \text{sign}(\overline{\omega_m}) \tag{3.10}$$

k_{mi} is the gain for the SM-MIO, and in order to maintain the SM observer stable, it should be tuned according to the Lyapunov stability criterion. Based on the SM theory, when the system gets to the equilibrium state, the real-time mutual inductance equals the estimated value, that is,

$$M_{sf} = M_{sf}^{est} = k_{mi} F(\overline{\omega_m}) \tag{3.11}$$

where M_{sf}^{est} represents the estimated mutual inductance. The block diagram of the SM-MIO is shown in Fig. 3.3. What needs to be mentioned is that firstly, when using the MIO, because L_d and L_q have not been observed up to now, the inductance values on nameplate (L_d^{mea} and L_q^{mea}) require to be employed in (3.9). Secondly, considering

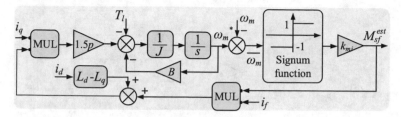

Fig. 3.3 Block diagram of the proposed mutual inductance observer

the inherent property (chattering effect) of an SM observer, a low-pass filter should be employed for a more accurate output.

(b) QAIO

The SM-QAIO based on (3.4) is described as follows:

$$\frac{di_d^*}{dt} = -\frac{R_s}{L_d}i_d^* + \frac{k_{qa}F(\overline{i_d})}{L_d}p\omega_m i_q - \frac{M_{sf}}{L_d}\frac{di_f}{dt} + \frac{1}{L_d}u_d \tag{3.12}$$

where i_d^* is the estimated d-axis current. $\overline{i_d}$ is the error between the estimated current and the real current. k_{qa} is the gain for the QAIO and $F(\overline{i_d})$ is the signum function.

$$F(\overline{i_d}) = \text{sign}(\overline{i_d}) \tag{3.13}$$

Similar as the MIO, when the system works stably, the real-time q-axis inductance will equal the estimated value, namely,

$$L_q = L_q^{est} = k_{qa}F(\overline{i_d}) \tag{3.14}$$

where L_q^{est} is the estimated q-axis stator inductance. The block diagram of the observer is shown in Fig. 3.4. Unlike the SM-MIO, because the mutual inductance has been obtained in the first step, it can be directly substituted into (3.12) for observer construction. Whereas, as far as the stator resistance and the d-axis inductance are concerned, the offline-tested values L_d^{mea} and R_s^{mea} are needed as well.

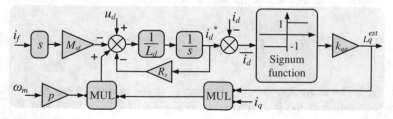

Fig. 3.4 Block diagram of the proposed q-axis inductance observer

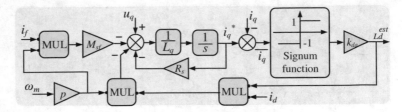

Fig. 3.5 Block diagram of the proposed d-axis inductance observer

(c)   DAIO

After detecting the mutual inductance and q-axis stator inductance, they can be directly employed to construct the DAIO on the basis of (3.2), which can be described as:

$$\frac{di_q^*}{dt} = -\frac{k_{da}F(\bar{i_q})}{L_q}p\omega_m i_d - \frac{R_s}{L_q}i_q^* - \frac{M_{sf}}{L_q}p\omega_m i_f + \frac{1}{L_q}u_q \qquad (3.15)$$

where i_q^* is the estimated q-axis current. $\bar{i_q}$ is the error between the estimated current and the real current. k_{da} means the gain for the DAIO and $F(\bar{i_q}) = \text{sign}(\bar{i_q})$. The block diagram of the observer is shown in Fig. 3.5. When the system arrives at the stable status, the d-axis inductance can be diagnosed as:

$$L_d = L_d^{est} = k_{da}F(\bar{i_q}) \qquad (3.16)$$

where L_d^{est} is the estimated q-axis stator inductance.

(d)   FWIO

In contrast with the differential Eqs. (3.1), (3.2) and (3.8), the field winding inductance information exists in every term of (3.3) in the form of denominators. These place great challenges on designing an SM observer. In detail, if the observer is constructed by just replacing all terms concerning L_{ff} with an estimated one in (3.3), like the previous SM observers, it is inclined to be unstable during operations. To solve this issue, we can use the offline-tested inductance L_{ff}^{mea} to reconstruct the SM observer for field winding inductance as follows:

$$\frac{di_f^*}{dt} = -\frac{3M_{sf}}{2L_{ff}^{mea}}\frac{di_d}{dt} - \frac{R_f}{L_{ff}^{mea}}i_f^* + k_f F(\bar{i_f})u_f \qquad (3.17)$$

where i_f^* is the estimated exciting current. $\bar{i_f}$ is the error between the estimated exciting current and the real current. k_f is the gain for the FWIO and $F(\bar{i_f}) = \text{sign}(\bar{i_f})$. Figure 3.6 illustrates the block diagram of the observer. M_{sf} is the value detected from the first step. By using the observer, the following result is expected

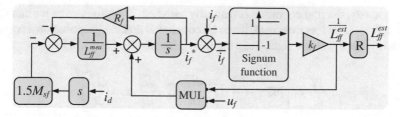

Fig. 3.6 Block diagram of the field winding inductance observer

to be obtained:

$$L_{ff} = L_{ff}^{est} = \frac{1}{k_f F(\overline{i_f})} \tag{3.18}$$

where L_{ff}^{est} is the estimated field winding inductance. Finally, it needs to be mentioned that the offline-tested rotor resistance R_f^{mea} should be also adopted in (3.17).

3.3.2 Stability Analysis

To analyze the stability of the proposed sliding mode observers, the Lyapunov function is an effective technique [20]. Considering that the proposed observers are implemented step by step by just substituting the outputs of the previous observers into the latter ones (see Fig. 3.1), they are independent intrinsically, so four different Lyapunov functions are needed to analyse the stability of each observer. Following the Lyapunov principle, the defined speed, d-axis, q-axis and exciting current sliding surfaces ($S_{\omega m}, S_d, S_q, S_f$) are $\overline{\omega_m}, \overline{i_d}, \overline{i_q}$ and $\overline{i_f}$, respectively. They can be denoted as:

$$\mathbf{S} = \left[S_{\omega_m}, S_d, S_q, S_f \right]^T = \left[\overline{\omega_m}, \overline{i_d}, \overline{i_q}, \overline{i_f} \right]^T \tag{3.19}$$

To analyze the stability of the developed observers, the following Lyapunov equations are constructed:

$$\mathbf{V} = diag(\frac{1}{2}\mathbf{S} \cdot \mathbf{S}^{\mathrm{T}}) = \frac{1}{2}\left[\overline{\omega_m}^2, \overline{i_d}^2, \overline{i_q}^2, \overline{i_f}^2 \right]^T \tag{3.20}$$

where $diag()$ represents the function used to extract diagonal elements of a matrix. In order to ensure the stability of the proposed observers, the following inequality conditions should be satisfied:

$$\mathbf{V} > 0, \quad \frac{d\mathbf{V}}{dt} < 0 \tag{3.21}$$

Obviously, the first condition is naturally satisfied. Therefore, once we can make certain that $\frac{dV}{dt}$ is less than zero, it can be concluded that the proposed observers are able to arrive at a stable state. Take the time derivative of Eq. (3.21):

$$\frac{dV}{dt} = \left[\overline{\omega_m} \cdot \frac{d\overline{\omega_m}}{dt}, \overline{i_d} \cdot \frac{d\overline{i_d}}{dt}, \overline{i_q} \cdot \frac{d\overline{i_q}}{dt}, \overline{i_f} \cdot \frac{d\overline{i_f}}{dt} \right]^T \tag{3.22}$$

Substitute (3.1–3.3), (3.8), (3.9), (3.12), (3.15) and (3.17) into (3.22), and it can be deduced that:

$$\frac{dV}{dt} = \begin{bmatrix} \underbrace{-\frac{B}{J}\overline{\omega_m}^2}_{term1} + \underbrace{\frac{3p}{2J}(k_{mi}F(\overline{\omega_m}) - M_{sf})i_f i_q \overline{\omega_m}}_{term2} \\[2mm] \underbrace{-\frac{R_s}{L_d}\overline{i_d}^2}_{term1} + \underbrace{\frac{k_{qa}F(\overline{i_d}) - L_q}{L_d}p\omega_m i_q \overline{i_d}}_{term2} \\[2mm] \underbrace{-\frac{R_s}{L_q}\overline{i_q}^2}_{term1} + \underbrace{\frac{-k_{da}F(\overline{i_q}) + L_d}{L_q}p\omega_m i_d \overline{i_q}}_{term2} \\[2mm] \underbrace{-\frac{R_f}{L_{ff}^{mea}}\overline{i_f}^2}_{term1} + \underbrace{(k_f F(\overline{i_f}) - \frac{1}{L_{ff}})u_f \overline{i_f}}_{term2} \end{bmatrix} \tag{3.23}$$

Obviously, *term*1 of the four equations is less than zero. In order to keep the SM observers being stable constantly, *term*2 of each equation is expected to be less than zero as well, that is,

$$\begin{cases} (k_{mi}F(\overline{\omega_m}) - M_{sf})i_f i_q \overline{\omega_m} < 0 \\[2mm] \dfrac{k_{qa}F(\overline{i_d}) - L_q}{L_d}p\omega_m i_q \overline{i_d} < 0 \\[2mm] \dfrac{-k_{da}F(\overline{i_q}) + L_d}{L_q}p\omega_m i_d \overline{i_q} < 0 \\[2mm] (k_f F(\overline{i_f}) - \dfrac{1}{L_{ff}})u_f \overline{i_f} < 0 \end{cases} \tag{3.24}$$

The above equations can be future derived as:

$$\begin{cases} (k_{mi} - M_{sf})i_f i_q < 0, & \text{if } \overline{\omega_m} > 0 \\ (-k_{mi} - M_{sf})i_f i_q > 0, & \text{if } \overline{\omega_m} < 0 \end{cases} \tag{3.25}$$

Table 3.1 SM-MIO stability conditions considering state values

States			k_{mi}
$\omega_m > 0$	$i_q > 0$	$i_d \geq 0$	$< -M_{sf}$
		$i_d < 0$	
	$i_q < 0$	$i_d \geq 0$	$> M_{sf}$
		$i_d < 0$	
$\omega_m < 0$	$i_q > 0$	$i_d \geq 0$	$< -M_{sf}$
		$i_d < 0$	
	$i_q < 0$	$i_d \geq 0$	$> M_{sf}$
		$i_d < 0$	

$$\begin{cases} (k_{qa} - L_q)\omega_m i_q < 0, & \text{if } \overline{i_d} > 0 \\ (-k_{qa} - L_q)\omega_m i_q > 0, & \text{if } \overline{i_d} < 0 \end{cases} \tag{3.26}$$

$$\begin{cases} (-k_{da} + L_d)\omega_m i_d < 0, & \text{if } \overline{i_q} > 0 \\ (k_{da} + L_d)\omega_m i_d > 0, & \text{if } \overline{i_q} < 0 \end{cases} \tag{3.27}$$

$$\begin{cases} (k_f - \frac{1}{L_{ff}})u_f < 0, & \text{if } \overline{i_f} > 0 \\ (-k_f - \frac{1}{L_{ff}})u_f > 0, & \text{if } \overline{i_f} < 0 \end{cases} \tag{3.28}$$

From (3.25)–(3.28), it can be noted that apart from the errors between the estimated and the real values, the state variables (speed, exciting current, d, q-axis armature currents and exciting voltage) also influence the stability of the observers. In practice, when a WFSM works, i_f and u_f usually stand at the positive position, but i_q, i_d and ω_m might not be fixed, increasing the difficulty in determining the stability conditions. However, this problem can be easily solved by implementing the observers in particular conditions. For example, when a motor rotates anticlockwise under load using a flux-weakening control strategy, i_q and ω_m can be controlled to remain positive while i_d is negative. Tables 3.1, 3.2, 3.3 and 3.4 show the stability conditions of each observer, which comprehensively consider the status of the variables.

Table 3.2 QAIO stability conditions considering state values

States			k_{qa}
$\omega_m > 0$	$i_q > 0$	$i_d \geq 0$	$< -L_q$
		$i_d < 0$	
	$i_q < 0$	$i_d \geq 0$	$> L_q$
		$i_d < 0$	
$\omega_m < 0$	$i_q > 0$	$i_d \geq 0$	$> L_q$
		$i_d < 0$	
	$i_q < 0$	$i_d \geq 0$	$< -L_q$
		$i_d < 0$	

Table 3.3 DAIO stability conditions considering state values

States			k_{da}
$\omega_m > 0$	$i_q \geq 0$	$i_d > 0$	$> L_d$
		$i_d < 0$	$< -L_d$
	$i_q < 0$	$i_d > 0$	$> L_d$
		$i_d < 0$	$< -L_d$
$\omega_m < 0$	$i_q \geq 0$	$i_d > 0$	$< -L_d$
		$i_d < 0$	$> L_d$
	$i_q < 0$	$i_d > 0$	$< -L_d$
		$i_d < 0$	$> L_d$

Table 3.4 FWIO stability conditions considering state values

States			k_f
$\omega_m > 0$	$i_q \geq 0$	$i_d \geq 0$	$< -\frac{1}{L_{ff}}$
		$i_d < 0$	
	$i_q < 0$	$i_d \geq 0$	
		$i_d < 0$	
$\omega_m < 0$	$i_q \geq 0$	$i_d \geq 0$	
		$i_d < 0$	
	$i_q < 0$	$i_d \geq 0$	
		$i_d < 0$	

Definitely, once both the machine working status and the gain coefficients are consistent with the information in Tables 3.1, 3.2, 3.3 and 3.4, all of the proposed observers will remain stable when they are implemented one by one. Before leaving these tables, several interesting phenomena should be addressed. Firstly, it can be seen from Tables 3.1, 3.2 and 3.4 that the state of d-axis armature current does not influence the stability of the observers (3.9), (3.12) and (3.17) at all, and i_d can be set as zero (rather than just the negative and positive values). But for the DAIO, i_d cannot be controlled to maintain at zero. Secondly, no-load operations ($i_q = 0$) are only allowed for the DAIO and FWIO, but when the SM-MIO and QAIO are implemented, this condition should be avoided. Specifically, the next section provides a way for determining the working states, which can reject the impacts of parameter uncertainties.

3.3.3 Observer Robustness Against Parameter Uncertainties

Except the DAIO, the other three observers are achieved by using the measured inductance values. Whereas, because the offline parameters might be different from the real ones when a WFSM works (inductance mismatch phenomenon), the accuracy of the proposed inductance observers might degrade thereout. Besides, the measured resistances and mechanical parameters are also main factors influencing the accuracy of the observers. To solve the problem, the robustness of the proposed observers against parameter mismatch should be analyzed.

(a)   SM-MIO

Substitute L_d^{mea} and L_q^{mea} into (3.9) and subtract (3.8) from (3.9), and the mutual inductance estimation error can be obtained:

$$err\_mi = M_{sf}^{est} - M_{sf} = \underbrace{\frac{2J}{3pi_q i_f}[\frac{d(\omega_m^* - \omega_m)}{dt} + \frac{B}{J}(\omega_m^* - \omega_m)]}_{term1}$$

$$+ \underbrace{\frac{i_d}{i_f}[(L_q^{mea} - L_q) - (L_d^{mea} - L_d)]}_{term2} \qquad (3.29)$$

where err_mi is the estimation error for the mutual inductance. When the system reaches an equilibrium state, because the error between the estimated and the real-time speed is zero and the differential of it is zero, $term1$ equals zero. In this case, the impacts of the mechanical parameters J and B on the observation accuracy can also be rejected. Consequently, the mutual inductance estimation error equals $term2$, which is directly relevant to i_d. Intuitively, in order to ensure that the estimation error is able to approach zero, i_d should be controlled to near zero when implementing the SM-MIO.

(b)   QAIO

As for the QAIO, only the d-axis stator inductance is tested offline. When substituting L_d^{mea} into (3.12) and then subtracting (3.1) from it, the estimation error can be described as:

$$err\_qa = \frac{1}{p\omega_m i_q}[L_d^{mea}\frac{di_d^*}{dt} - L_d\frac{di_d}{dt} + R_s(i_d^* - i_d)] \qquad (3.30)$$

where err_qa is the estimation error for q-axis armature inductance. Considering that $i_d^* - i_d = 0$ and $\frac{di_d^*}{dt} = \frac{di_d}{dt}$ when the system becomes stable, it can be further derived that:

$$err\_qa = \frac{(L_d^{mea} - L_d)}{p\omega_m i_q}\frac{di_d}{dt} \qquad (3.31)$$

It should be noticed that there is only one way to make err_qa zero constantly, that is, controlling i_d to stabilize at a constant value. Besides, it can be seen that the impacts of R_s^{mea} can be avoided due to the nature of the observer, which is the same as the DAIO of which structure is similar to that of the QAIO.

(b)   FWIO

With reference to the aforementioned circumstances, by subtracting (3.3) from (3.17), the estimation error for the reciprocal field winding inductance $err_{\_fw}$ can be deduced as:

$$err_{\_fw} = \frac{1}{L^{est}_{ff}} - \frac{1}{L_{ff}} = \frac{d(i^*_f - i_f)}{dt} \cdot \left(u_f - \frac{3M_{sf}}{2u_f}\frac{di_d}{dt} - i_f R_f\right) \qquad (3.32)$$

It can be noted that once the estimated exciting current converges to the real one (stable state), the field winding inductance estimation error is zero regardless of the working states. Hence, it can be concluded that the FWIO has strong robustness against parameter variations.

When implementing the SM-MIO, apart from the stability conditions, the d-axis reference current should be near zero during control for the sake of a relatively high diagnostic accuracy. As for the QAIO, the d-axis current needs to level off when the algorithm is executed. Comparatively speaking, no special working statuses are required when the FWIO algorithms are implemented. In terms of the DAIO, due to the stability condition (3.27), the d-axis current cannot be controlled to maintain at zero (either larger or less than zero), but this does not influence its robustness at all. In summary, except the FWIO, the other three observers have high estimation accuracy only in the particular conditions (relying on the d-axis current). Hence, it needs to be mentioned that for the sake of high accuracy considering all of the four observers, the proposed multistep inductance identification strategy is more applicable to the low-d-axis current applications.

3.4   Implementations of SM-Observer-Based FCS-MPC

Based on the block diagram of the FCS-MPCC structure in Fig. 3.1. At the kth period, the proposed SM-Observer-based FCS-MPCC implementation procedures are:

(a)   Measurement: detect the real-time phase currents, rotor position, torque and speed by using sensors, and the excitation current is estimated based on the Kirchhoff's law.

(b)   *abc/dq* transformation: the measured phase currents are transformed to the d, q-axis currents according to the rotor position.

(c)   Observation: use the immediate currents, voltages to calculate the mutual inductance, armature inductances and self-inductance of the field windings by using the SM observers.

(d)   Prediction: use the immediate speed, currents and inductance parameters to estimate the future current states $i_d(k + 1)$ and $i_q(k + 1)$ for all the seven candidate voltage vectors (see (2.13)) by using (3.6) and (3.7).

(e)   Evaluation: substitute the seven the predicted values into the cost function (3.33) and determine the optimal voltage vector and the corresponding switching states.

$$J = \left|i_{d\_ref} - i_d(k+1)\right| + \left|i_{q\_ref} - i_q(k+1)\right| \tag{3.33}$$

where i_{d_ref} and i_{q_ref} are the reference currents.

(f) Actuation: apply the optimum switching state to the drive system.

3.5 Verifications

The proposed multistep inductance identification methods and the proposed inductance observer-based FCS-MPCC are validated by simulation in this section. The parameters of the WFSM prototype, which are provided by the manufacturer, and the control parameters of the system are consistent with Table 3.5. It deserves to be mentioned that the manufacturer measured the machine parameters by using an offline identification method which does not need to consider the magnetic saturation effect. The verifications are divided into two parts: parameter identification discussion and FCS-MPCC discussion. It deserves to be mentioned that in the first part, the FCS-MPCC strategy that uses the fixed inductance values (measured values) is adopted for machine control, while the parameters of the SM observer can be changed to simulate the parameter mismatch situations. In the second part, the performance characteristics of the FCS-MPCC methods without inductance mismatch, with inductance mismatch and by using the real-time parameters are compared.

Table 3.5 Motor and control parameters

Parameter	Value	Unit
stator winding resistance R_s	2.74	Ω
rotor winding resistance R_f	19.5	Ω
d-axis inductance L_d	150	mH
q-axis inductance L_q	130	mH
field winding inductance L_{ff}	2.97	H
mutual inductance M_{sf}	0.78	H
exciting voltage u_f	37.4	V
the number of pole pairs p	2	–
moment of inertia J	0.08	kg·m^2
damping coefficient B	0.0012	–
maximum speed ω_{m_max}	73	rad/s
DC-link voltage U_{dc}	400	V
rated torque T_{rated}	8.0	Nm

3.5.1 Parameter Identification Results

For the sake of a comprehensive discussion, the following simulation results are
included in this section. Firstly, the effectiveness of the proposed inductance identification methods is verified in both low-speed and high-speed conditions. Simultaneously, the detailed implementation procedures of the multistep observers are
demonstrated. Secondly, the robustness of the SM observers, which is explained in
Sect. 3.3.3, is analyzed. In simulation, the parameter values in Table 3.5 are regarded
as the real ones.

(a)   Effectiveness

Considering that the estimated results of an SM observer are more credible in the
stable states (illustrated in the last part), in which no external disturbances (e.g.,
speed and torque fluctuations) emerge, the estimated results of the proposed inductance observers are analyzed when the machine runs stably in simulation. The verification procedures for the proposed inductance estimation techniques are designed
as follows. Firstly, the machine speed is controlled to maintain at 21 rad/s (low
speed) and 73 rad/s (high speed), respectively, under the rated load (8 Nm) for 3 s
(t_0), in which d-axis current reference is set as zero. In this case, the real-time
mutual inductance M_{sf}^{est} is recorded and its average value is calculated. Secondly, the
average (ave) value of M_{sf}^{est} is substituted into the QAIO, and simultaneously, the d-
axis current reference is set as -2 A without changing the speed and load conditions
($\omega_m > 0$, $i_q > 0$). At this stage, the q-axis inductance L_q^{est} is obtained. Thirdly, keep
the test conditions the same to those in the second step and substitute the previously
estimated values into the DAIO and FWIO, calculating the values of L_d^{est} and L_{ff}^{est}
sequentially.

Figure 3.7 shows the simulation results of control performance when implementing the proposed SM observers at the speed of 21 rad/s. As for the SM-MIO,
firstly, it can be seen that the machine speed levels off at 21 rad/s, and the d- and
q-axis currents are zero and -2 A with slight fluctuations in Fig. 3.7a, respectively.
It deserves to be mentioned that k_{mi} is set to be smaller than $-M_{sf}$ in this situation.
Secondly, when the SM-MIO reaches the equilibrium state, the estimated rotor speed
is able to track the actual speed well, and the error between them is zero. Thirdly, the
estimated average mutual inductance is 0.78 H, complying with the offline measured
value. This represents that the mutual inductance observer based on SM theory is
effective. When implementing the QAIO algorithm (see Fig. 3.7b), the speed and
q-axis current are over zero (21 rad/s and 2 A, respectively), while the d-axis current
is about -2 A (negative). Hence, according to Table 3.2, k_{qa} satisfies the following
equation: $k_{qa} < -L_q$. Because the estimated d-axis current of the QAIO is able to
track the real value accurately ($\widetilde{i_d} = 0$), the observer is stable. Definitely, the predicted
q-axis inductance equals the offline value, representing that the proposed QAIO is
effective in the low-speed conditions. In Fig. 3.7c, the performance characteristics
of currents and speed are the same to those in Fig. 3.7b. In this case, k_{da} is set to be
smaller than $-L_d$ in the simulation. As for the DAIO, the error between the estimated

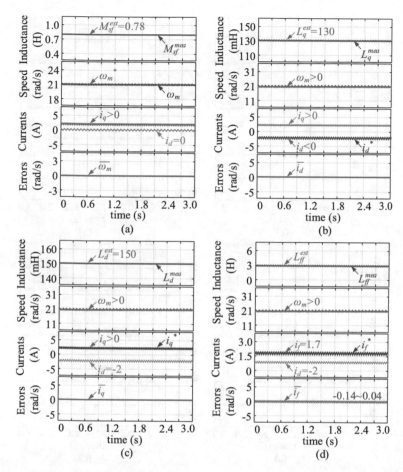

Fig. 3.7 Simulation results at the speed of 21 rad/s. **a** Performance when SM-MIO is implemented. **b** Performance when QAIO is implemented. **c** Performance when DAIO is implemented. **d** Performance when FWIO is implemented

q-axis current and the real current is zero, and the estimated d-axis inductance is 150 mH, being in accordance with the off-line value. Theoretically, the proposed DAIO is effective. As for the FWIO, k_f has nothing to do with the test conditions and it must be set to be smaller than $-1/L_{ff}$. It can be noted that the error between the estimated exciting current (nearly 1.7 A) and the real current is small, though it does not remain at zero like the other observers. In Fig. 3.7d, it can be seen that the estimated field winding inductance is about 2.97 H, proving that the proposed FWIO is effective over the low-speed range.

Figure 3.8 illustrates the performance of the system when the rotor speed is 73 rad/s. First of all, the machine speed can level off at the reference level when the proposed observers are implemented. This lays the foundations for achieving effective observations. Then, the errors between the estimated states ($\omega_m{}^*$, $i_d{}^*$, $i_q{}^*$

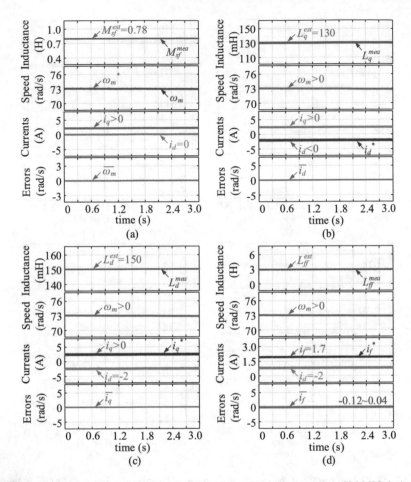

Fig. 3.8 Simulation results at the speed of 73 rad/s. **a** Performance when SM-MIO is implemented. **b** Performance when QAIO is implemented. **c** Performance when DAIO is implemented. **d** Performance when FWIO is implemented

and $i_f{}^*$) in the four observers and the actual values are nearly zero, which is similar to the results in Fig. 3.7. Hence, the proposed observers are able to get stable during control once the gain coefficients satisfy the conditions in Table 3.1, 3.2, 3.3 and 3.4. Most importantly, the observed mutual inductance, q-axis inductance, d-axis inductance and field winding inductance are 0.78 H, 130 mH, 150 mH and 2.97 H, respectively, which are consistent with the values measured offline. Above all, the simulation results prove that the proposed multistep inductance identification strategy is effective regardless of the magnitude of machine speed.

According to the above analysis, an interesting phenomenon that should not be ignored is that the estimated mutual inductance at the zero-d-axis current condition does not greatly influence the accuracy of the other observers working at the -2 A

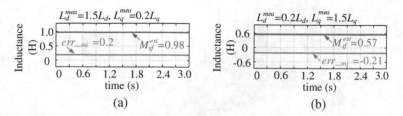

Fig. 3.9 Mutual inductance estimation when parameter mismatch occurs. **a** $L_d^{mea} = 1.5L_d, L_q^{mea} = 0.2L_q$. **b** $L_d^{mea} = 0.2L_d, L_q^{mea} = 1.5L_q$

condition. This proves that the proposed multistep inductance estimation strategy has high accuracy simultaneously as long as the d-axis current is low.

(b) Robustness

According to Sect. 3.3.3, three phenomena need to be verified. Firstly, when inductance mismatch problem occurs, the accuracy of the proposed SM-MIO, QAIO and FWIO will degrade if the test conditions do not comply with the given ones. Then, the inductance estimation errors are derived theoretically, but the accuracy of them should be discussed. Finally, the conditions used for ensuring the robustness of the observers are effective.

In terms of the SM-MIO, assume that the measured d, q-axis inductances deviate from the real values, and the test conditions include that the machine speed is 73 rad/s under 8 Nm and the d-axis current reference is set as -2 A. Figure 3.9a, b show the estimated mutual inductance when the measured d, q-axis inductances are assumed to satisfy that $L_d^{mea} = 1.5L_d$, $L_q^{mea} = 0.2L_q$ and $L_d^{mea} = 0.2L_d, L_q^{mea} = 1.5L_q$, respectively. It can be noted that when the d, q-axis inductance mismatch phenomenon occurs, although the observer can get stable, the estimated mutual inductance cannot reflect the real values. Specifically, in Fig. 3.9a, the output of the SM-MIO is 0.98 H while the inductance estimation error is 0.2 H, and in Fig. 3.9b, they are 0.57 H and -0.21 H, respectively. It deserves to be mentioned that the results are consistent with the theoretical results in (3.29). Figure 3.9 intuitively proves that the accuracy of the SM-MIO will be affected by the measured d, q-axis inductances, and the theoretical inductance estimation errors are verified to be correct to some extent. However, they are not sufficient. To solve the problem, more simulation results are provided. In detail, Fig. 3.10 illustrates the impacts of the d-axis current and the assumptive measured d, q-axis inductances on the estimation errors for the mutual inductance err_{mi}. It can be seen that, firstly, when the assumptive inductance remains at the same level, the larger the magnitude of the d-axis current is, the larger err_{mi} becomes. And if the measured inductances are larger than the actual values, err_{mi} is negative. Otherwise, it is positive. Secondly, the mutual inductance estimation errors obtained by simulation are in line with the theoretical values calculated by (3.29). Thirdly, when the d-axis current is near zero, the estimation error is small regardless of the values of the measured inductances. Therefore, for the sake of a relatively

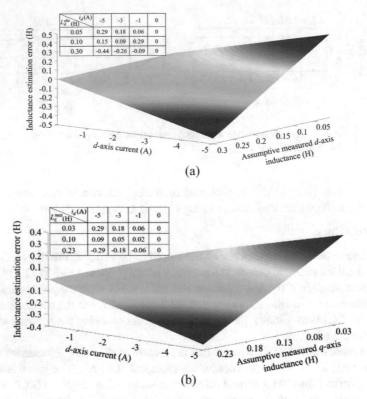

L_d^{obs} (H) \ i_d(A)	-5	-3	-1	0
0.05	0.29	0.18	0.06	0
0.4 / 0.10	0.15	0.09	0.29	0
0.3 / 0.30	-0.44	-0.26	-0.09	0

(a)

L_q^{mea} (H) \ i_d(A)	-5	-3	-1	0
0.03	0.29	0.18	0.06	0
0.10	0.09	0.05	0.02	0
0.23	-0.29	-0.18	-0.06	0

(b)

Fig. 3.10 Impacts of d-axis current and assumptive measured d, q-axis inductances on mutual inductance estimation errors. **a** Mismatched d-axis inductance. **b** Mismatched q-axis inductance

strong robustness, it is reasonable to control i_d to maintain at a low position when implementing the SM-MIO algorithm.

As far as the QAIO is concerned, (3.31) indicates that the estimation accuracy of the q-axis inductance is closely related to L_d^{mea} and the transient behaviors of i_d. In order to explain the problem explicitly, the verification procedures are designed as follows. The machine is controlled to rotate at 73 rad/s under rated load. Then, the d-axis current reference is -2 A between 0 and 0.15 s, but it starts to fall at 200 A/s afterward. During simulation, assume that the measured d-axis inductance is mismatched and $L_d^{mea} = 0.2L_d$. The simulation results are shown in Fig. 3.11. Firstly, when i_d levels off, even though the d-axis inductance does not comply with the real value, the inductance estimation result is accurate. Secondly, when i_d change in a fixed slope, the QAIO is still stable. However, the estimated q-axis inductance rises to nearly 200 mH, and the estimation error is about 70 mH. These prove that the proposed condition (i_d should be constant) used for ensuring robustness in this part is effective. Thirdly, when substituting the test conditions (di_d/dt = -200, ω_m = 73 rad/s) into (3.31), the calculated estimation error is 73 mH, which is very close to the simulation result. Thus, the theoretical analysis is pretty accurate. Finally, it

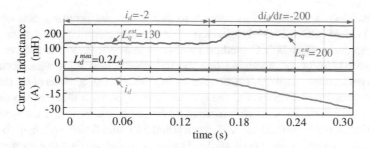

Fig. 3.11 q-axis inductance estimation results when $L_d^{mea} = 0.2L_d$

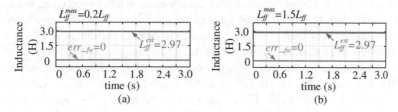

Fig. 3.12 Field winding inductance estimation when parameter mismatch occurs. **a** $L_{ff}^{mea} = 0.2L_{ff}$. **b** $L_{ff}^{mea} = 1.5L_{ff}$

needs to be mentioned that when i_d changes, the system must not be stable. As is explained before, the observers should work in the stable states, leading to the fact that the influence of mismatched d-axis inductance on QAIO is slight. Moreover, for the sake of intuitiveness, the slope of the d-axis current is set to be large. However, in practice, the changing rate of the current might be much smaller than -200 A/s considering safety, so the proposed QAIO has strong robustness against mismatched d-axis inductance actually even in the transient process.

Equation (3.17) indicates that the measured field winding inductance needs to be used for the FWIO. To discuss the impact of it on the inductance estimation accuracy, assume that $L_{ff}^{mea} = 0.2L_{ff}$ and $L_{ff}^{mea} = 1.5L_{ff}$, respectively, when the FWIO algorithms are implemented at 73 rad/s and 8 Nm ($i_d= -2$ A), and the simulation results are demonstrated in Fig. 3.12. Interestingly, the estimated field winding inductance can remain at nearly 2.97 H even though the parameter mismatch issue occurs. This represents that the proposed FWIO has pretty strong robustness in practice.

3.5.2 FCS-MPCC Control Results

It has been proven that the proposed inductance observers are suitable for the low-d-axis current situations in Sect. 3.5.1. On this basis, the simulation procedures and conditions are designed as follows to verify that the proposed SM-observer-based FCS-MPCC method is effective. Firstly, the control performance of the fixed-parameter FCS-MPCC method (without mismatch issue for the controller) is presented. Secondly, assuming that the inductance mismatch issue occurs, the control performance is given. Thirdly, the performance characteristics of the proposed SM-observer-based FCS-MPCC method are compared with those in the previous two cases when the inductance mismatch problem occurs. Fourthly, the machine is controlled to speed up to 21 rad/s between 0 and 5 s under 8 Nm, and then, the speed reference is set as 73 rad/s until 10.0 s. During test, the d-axis current reference is set as -0.5 A (low d-axis current).

Figure 3.13 shows the speed, torque and current performance when the traditional FCS-MPCC controller encounters no mismatch issue. It can be seen that, firstly, the speed tracking performance is good, which means that the machine speed can tack the reference values well. Secondly, the d-axis current maintains at nearly -0.5 A. Although large current ripples can be seen, the signal after filtering is used for observer construction. Hence, it does not influence the stability of the observers. Thirdly, the q-axis and torque surges at 5.0 s are 15.4 A and 63 A, respectively. These represent that the FCS-MPCC strategy is suitable for controlling the WFSMs.

In Fig. 3.14, the control performance of the traditional FCS-MPCC method with parameter mismatch (the d-axis, q-axis and mutual inductances used to construct the controller are 2, 0.5 and 1.5 times of the measured values, respectively) is illustrated. In comparison with Fig. 3.13, the performance decreases visibly. In detail, the speed and d-axis current cannot track the reference values in the high-speed situations. This

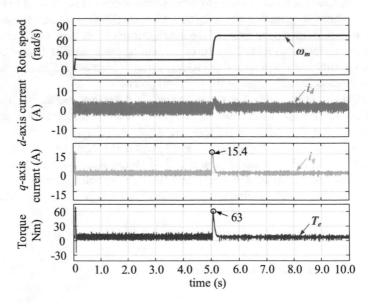

Fig. 3.13 Control performance when traditional FCS-MPCC controller encounters no mismatch issue

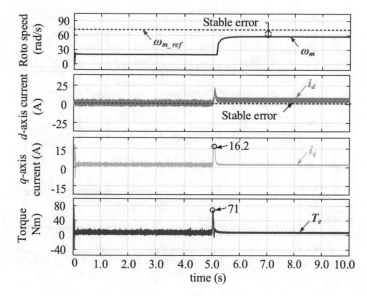

Fig. 3.14 Control performance when traditional FCS-MPCC controller encounters mismatch issue

demonstrates that the parameter mismatch issue influences the tracking performance of the drive system. Besides, the q-axis and torque surges at 5.0 s are slightly larger than those in Fig. 3.13, indicating that when parameter mismatch issue occurs, the dynamics of the system can be influenced. Overall, for the traditional FCS-MPCC method, if the parameters of the controller mismatch, the control performance will degrade inevitably.

Figure 3.15 depicts the system performance when the proposed SM-observer-based FCS-MPCC method is implemented in the parameter mismatch situation (the d-axis, q-axis and mutual inductances used to construct the controller are 2, 0.5 and 1.5 times of the measured values, respectively). Compared to Figs. 3.13 and 3.14, there are several features that need to be addressed in Fig. 3.15. Specifically, the speed and d-axis current can track the references, and the q-axis and torque surges at 5.0 s are close to those in Fig. 3.13. These prove that the proposed SM-observer-based FCS-MPCC method has relatively strong robustness against parameter mismatch.

3.6   Summary

This chapter proposes a robust SM observer-based multistep inductance (mutual inductance, d, q-axis inductances and field winding inductance) identification strategy to improve the robustness of the FCS-MPCC method used in the novel WFSM drives against inductance mismatch. The main contributions of this chapter are as follows:

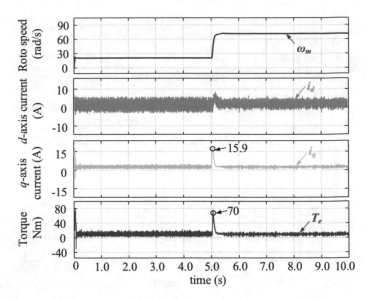

Fig. 3.15 Control performance of the proposed SM-observer-based FCS-MPCC controller

(1)   Based on the mathematical model (four different equations) of a WFSM, the mutual inductance observer, d, q-axis inductance observers and field winding inductance observer based on SM theory are established. Meanwhile, the sequential implementation procedures of the proposed observers are introduced. Because there have been few studies focusing on the SM principle-based inductance estimation techniques, especially for the novel WFSMs, the contents are novel as well as valuable.

(2)   By using the Lyapunov functions and numerical analysis, the gain factors in the observers and the machine operating conditions are designed to ensure the stability and robustness of the SM observers. The simulation results have verified the effectiveness, stability and robustness of the proposed observers. Because the proposed identification strategy is able to accurately detect the machine inductances regardless of magnetic saturation, it is highly valued in engineering.

(3)   Based on the machine model, an FCS-MPCC control method is derived for the capacitive coupling excitation-based WFSMs. SM inductance observers are innovatively integrated into the control process to provide accurate parameters used for robustness improvement.

Finally, it needs to be mentioned that apart from the inductances, the stator and rotor resistances are crucial parameters for the WFSMs, which might change greatly during operation. Hence, it is significant to investigate how to identify them accurately. The reason why this issue is not focused on in this part is that it is difficult to detect the machine resistances by using the pure SM theories because according to the mathematical model of the WFSMs, the resistances are directly linked with the state variables constructing the differential terms (stability conditions cannot be obtained). This means novel resistance identification methods need to be developed for the WFSMs in the near future. Although the resistances are not observed in this study, their impacts on the inductance estimation accuracy are discussed and the working conditions used for avoiding them are given, which are crucial for ensuring the robustness of the proposed observers. Another disadvantage of the proposed multistep inductance identification technique is that in order to ensure high accuracy for all of the four observers, the applications are limited to the low-d-axis current situations, which needs to be improved in the future.

References

1. G. Jawad, Q. Ali, T.A. Lipo, B. Kwon, Novel brushless wound rotor synchronous machine with zero-sequence third-harmonic field excitation. IEEE Trans. Magnet. **52**(7), 1–4 (2016). (Art no. 8106104)
2. F. Yao, Q. An, X. Gao, L. Sun, T.A. Lipo, Principle of operation and performance of a synchronous machine employing a new harmonic excitation scheme. IEEE Trans. Indus. Appl. **51**(5), 3890–3898 (2015)
3. D.C. Ludois, J.K. Reed, K. Hanson, Capacitive power transfer for rotor field current in synchronous machines. IEEE Trans. Power Electron. **27**(11), 4638–4645 (2012)
4. R. Gunabalan, P. Sanjeevikumar, F. Blaabjerg, O. Ojo, V. Subbiah, Analysis and implementation of parallel connected two-induction motor single-inverter drive by direct vector control for industrial application. IEEE Trans. Power Electron. **30**(12), 6472–6475 (2015)
5. M. Moradian, J. Soltani, A. Najjar-Khodabakhsh, G.R.A. Markadeh, Adaptive torque and flux control of sensorless IPMSM drive in the stator flux field oriented reference frame. IEEE Trans. Industr. Inf. **15**(1), 205–212 (2019)
6. S.A. Odhano, P. Pescetto, H.A.A. Awan, M. Hinkkanen, G. Pellegrino, R. Bojoi, Parameter identification and self-commissioning in ac motor drives: a technology status review. IEEE Trans. Power Electron. **34**(4), 3603–3614 (2019)
7. L. Peretti, M. Zigliotto, Automatic procedure for induction motor parameter estimation at standstill. IET Electr. Power Appl. **6**(4), 214–224 (2012)
8. T. Košt'ál, "Offline induction machine parameters identification suitable for self-commissioning," *2017 International Conference on Applied Electronics (AE)*, Pilsen, (2017), pp. 1–4
9. X. Wu, X. Fu, M. Lin, L. Jia, Offline inductance identification of IPMSM with sequence-pulse injection. IEEE Trans. Industr. Inf. **15**(11), 6127–6135 (2019)
10. Y. Ouyang, Y. Dou, "Speed sensorless control of PMSM based on MRAS parameter identification," *2018 21st International Conference on Electrical Machines and Systems (ICEMS)*, Jeju, (2018), pp. 1618–1622
11. S. Ichikawa, M. Tomita, S. Doki, S. Okuma, Sensorless control of permanent-magnet synchronous motors using online parameter identification based on system identification theory. IEEE Trans. Industr. Electron. **53**(2), 363–372 (2006)
12. Y. Shi, K. Sun, L. Huang, Y. Li, Online identification of permanent magnet flux based on extended kalman filter for IPMSM drive with position sensorless control. IEEE Trans. Industr. Electron. **59**(11), 4169–4178 (2012)
13. S. Wang, V. Dinavahi, J. Xiao, Multi-rate real-time model-based parameter estimation and state identification for induction motors. IET Electr. Power Appl. **7**(1), 77–86 (2013)
14. H. Bai, P. Zhang, V. Ajjarapu, A novel parameter identification approach via hybrid learning for aggregate load modeling. IEEE Trans. Power Syst. **24**(3), 1145–1154 (2009)

15. S.M.N. Hasan, I. Husain, A Luenberger–sliding mode observer for online parameter estimation and adaptation in high-performance induction motor drives. IEEE Trans. Indus. Appl. **45**(2), 772–781 (2009)
16. A. S. Tlili, E. B. Braiek, "A reduced-order robust observer using nonlinear parameter estimation for induction motors," *IEEE International Conference on Systems, Man and Cybernetics*, (Yasmine Hammamet, Tunisia, 2002), pp. 1–5
17. Zhang, Z. Zhao, T. Lu, L. Yuan, W. Xu, J. Zhu, "A comparative study of Luenberger observer, sliding mode observer and extended Kalman filter for sensorless vector control of induction motor drives," *2009 IEEE Energy Conversion Congress and Exposition*, (San Jose, CA, USA, 2009), pp. 2466–2473
18. C. Gong, Y. Hu, J. Gao, Y. Wang, L. Yan, An improved delay-suppressed sliding-mode observer for sensorless vector-controlled PMSM. IEEE Trans. Industr. Electron. **67**(7), 5913–5923 (2020)
19. Y. Zhou, "Research on direct torque control for electrically excited synchronous motors," Ph.D. dissertation, College of Automation Engineering, Nanjing University of Aeronautics and Astronautics, Nanjing, China, 2006
20. H. Kim, J. Son, J. Lee, A high-speed sliding-mode observer for the sensorless speed control of a PMSM. IEEE Trans. Industr. Electron. **58**(9), 4069–4077 (Sept. 2011)

Chapter 4
MPC Accuracy Improvement for PMSMs—Part I

Yaofei Han, Chao Gong⊚, and Jinqiu Gao

This chapter is divided into two parts, investigating how to improve the prediction accuracy of the finite control set model predictive control (FCS-MPCC) method. In Sect. 4.1, in order to enhance the control performance of the surface-mounted permanent magnet synchronous machines (PMSM) working under the low frequency situations, an accurate FCS-MPCC method is innovatively proposed. Firstly, a novel predicting plant model (PPM) in the continuous-time domain based on the numerical solutions of the PMSM state-space model (differential equations) is developed. Without using the linear discretization implementation, the influence of the low control frequency (LCF) can be eliminated completely. Besides, a brand-new calculation delay compensation method based on delay time prediction and current pre-compensation is designed for the proposed FCS-MPCC strategy. In Sect. 4.2, a new FCS-MPC strategy that simultaneously evaluates two targeting control objectives (TCOs) including speed and currents in a single cost function is proposed, achieving high-performance single-closed-loop control structure. Besides, aiming at the calculation delay problem of the MPC controllers, a novel calculation delay compensation method by predicting the current variation within the delay time is proposed. In this part, an improved machine model that is specially designed for the multi-objective FCS-MPC operation is illustrated at first. Then, a new cost function which can evaluate the tracking performance of speed and d-axis current and steady-state performance of q-axis current is developed. Compared to the conventional FCS-MPC approaches, extra speed controllers are not needed so that the proposed control topology becomes simpler. Then, in order to tune the weighting factors for the speed and currents in the cost function, an efficient handling strategy containing two implementation procedures, state variable normalization (SVN) and balance of

Y. Han
National Maglev Transportation Engineering R&D Center, Tongji University, Shanghai 201804, China

C. Gong (✉) · J. Gao
School of Automation, Northwestern Polytechnical University, Xi'an 710072, China

state sensitivity to voltage alteration, is developed. Finally, a brand-new computation delay estimation and compensation technique based on dual-sampling within a control period is proposed to reduce the current and torque ripples during control process.

4.1 Numerical Solution-Based FCS-MPCC

4.1.1 Problem Descriptions

Nowadays, most of the FCS-MPCC methods are achieved relying on the forward Euler discretization-based plant model of the machine, and usually, the discretization time step equals the sampling (control) cycle [1]. Practically, in order to reduce the torque and current ripples of the drive system, the control frequency should stand at a pretty high position (e.g., 10 kHz). As is shown in paper [2], the control performance of an FCS-MPCC controller is as remarkable as that of the space vector pulse width modulation (SVPWM) based field-oriented control (FOC) methods as the switching/control frequency is high. Whereas, it is widely acknowledged that the high switching/control frequency will generate quantities of loss and heat in the power devices, lowering the efficiency and reliability of the whole system. Considering this issue, many up-to-date researches have focused on the low control frequency (LCF) drives [3, 4]. However, the control performance will witness a marked degradation (e.g., higher torque and current ripples) when the system works at the LCF. As for the traditional FCS-MPCC strategy, one crucial reason for this phenomenon is that the discretization process is inaccurate. In detail, as shown in Fig. 4.1, the Euler discretization approach implies that the system state value i_p will shift in a linear trend when a particular candidate voltage vector is applied in each control period, while this default assumption does not conform to the real situations because of the nonlinear property of the system (real state is i_r). When the control period increases, the accuracy of the one-step prediction results will decline greatly. Namely, distinct

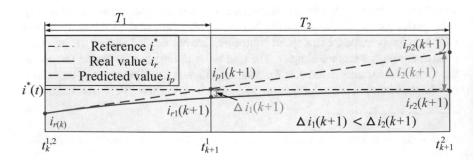

Fig. 4.1 Prediction process of traditional FCS-MPCC method

errors between the predicted values and the real ones will emerge as the control frequency becomes low. For example, in Fig. 4.1, when the control period increases from T_1 to T_2, the current error grows from $\triangle i_1(k + 1)$ to $\triangle i_2(k + 1)$. This will inevitably influence the optimal switching state selection process of an FCS-MPCC controller, and further degrade its control performance.

Another common problem of the FCS-MPCC strategy for PMSM drives is that the quantities of calculations have to be executed during the optimal control voltage process, resulting in that the calculation time is long [5, 6]. The time delay between the state measurement and actuation would lead to inaccurate selection of the control voltage and further deteriorate the system performance if it is not considered, lessening the effect of optimal control. Many scholars have addressed the time delay issue of FCS-MPC. For the traditional linear discretization-based FCS-MPCC, the most commonly used delay compensation method is the two-step prediction (TSP) strategy that uses the machine model shifted one step forward to calculate the manipulated voltages [7, 8]. These indicate that tackling the calculation problem is crucial for improving the control performance of an FCS-MPCC controller and it is highly required to develop effective delay compensation approaches for any newly developed FCS-MPCC methods.

In order to overcome the drawbacks of the traditional discretization method and increase the control performance for the LCF situations, a novel FCS-MPCC algorithm by using an innovative numerical solution based predicting plant model (PPM) is developed in this part. Without using the linear discretization, the traditional calculation delay compensation method based on TSP technique will no longer be totally applicable. On this ground a brand-new delay handling approach that includes two sequential procedures (dual-sampling-technique-based delay time estimation and current pre-compensation) is discussed. The experimental results verify the effectiveness of the proposed strategies.

4.1.2 Numerical Solution-Based Predicting Plant

In terms of FCS-MPCC, the targeting control objectives are the d, q-axis currents, so only the electrical equations are required for prediction. The electrical properties of a surface-mounted PMSM in the rotating reference frame can be illustrated as follows:

$$\begin{cases} \frac{di_d}{dt} = -\frac{R_s}{L_s}i_d + p\omega_m i_q + \frac{u_d}{L_s} \\ \frac{di_q}{dt} = -p\omega_m i_d - \frac{R_s}{L_s}i_q + \frac{u_q}{L_s} - \frac{\Psi_f}{L_s}p\omega_m \end{cases} \tag{4.1}$$

where ω_m is the rotor mechanical angular speed. i_d, i_q are the dq-axis currents. u_d, u_q are dq-axis control voltages. L_s is the stator inductance. R_s is the stator winding resistance. p and Ψ_f represent the number of pole pairs and the flux linkage, respectively. Instead of discretizing the machine model with a linear method, a novel PPM based on the numerical solutions of (4.1) will be established. To solve the differential equations, the solutions can be expressed as:

$$\begin{cases} i_d(t) = \frac{m}{c} + [\frac{y}{c}\sin(p\omega_m t) + \frac{z}{c}\cos(p\omega_m t)] \cdot \exp(-\frac{R_s t}{L_s}) \\ i_q(t) = -\frac{n}{c} + [\frac{y}{c}\cos(p\omega_m t) - \frac{z}{c}\sin(p\omega_m t)] \cdot \exp(-\frac{R_s t}{L_s}) \end{cases} \quad (4.2)$$

where

$$\begin{cases} m = -p^2 L_s \Psi_f \omega_m + p L_s \omega_m u_q + R_s u_d \\ n = p R_s \Psi_f \omega_m + p L_s \omega_m u_d - R_s u_q \\ c = p^2 L_s^2 \omega_m^2 + R_s^2 \\ y = c \cdot i_q(t_0) + n \\ z = c \cdot i_d(t_0) - m \end{cases} \quad (4.3)$$

where $i_d(t_0)$ and $i_q(t_0)$ are the boundary condition. During prediction, let $i_d(t_0)$ and $i_q(t_0)$ equal the instantaneous sampling values at the start of each control period. Then, the states at the next instant, $i_d(t_0 + T)$ and $i_q(t_0 + T)$, can be calculated by setting the time t as the control period T. Obviously, the proposed PPM can reflect the continuous (real) current shift trend in a machine, being able to obtain the more accurate prediction results in comparison with the traditional strategy, especially in the LCF applications.

Before leaving the PPM, it should be noticed that the numerical solution-based method is more complex due to the operations of the exponential and trigonometric functions compared to the traditional FCS-MPCC method [1], resulting in larger computation burden. In Sect. 4.1.3, the calculation delay caused by the proposed algorithms will be compensated using a brand-new technique.

4.1.3 Novel Calculation Delay Compensation

The calculation delay effect is illustrated in Fig. 4.2a, where i^* and i_r are the trajectory of reference current and real current, respectively. $v_1, \ldots, v_7$ represent the seven voltage candidates ($v_{000}, v_{100}, v_{110}, v_{010}, v_{011}, v_{001}, v_{101}$), which can be derived by (2.13) and (2.14). i_{p_n} is the estimated current corresponding to the different voltage vectors and i_s is the estimated current when the calculated voltage vector is applied. Because there are fixed algorithms to be implemented in each control period for a PMSM drive, the calculation time τ can be assumed to be identical for each cycle. It can be noted that between t_{k-1} and t_k, the voltage vector v_4 is the optimal manipulated voltage if the calculation delay is ignored. However, as a result of τ, the selected switching state is applied with delay at $t_{k-1} + \tau$, leading to that the current locus cannot be controlled as expected and a large deviation occurs after a control

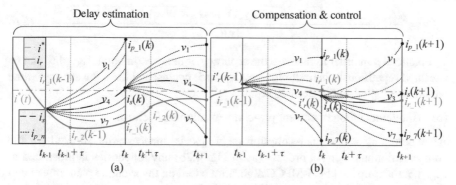

Fig. 4.2 Proposed delay compensation strategy. **a** Delay effect and delay prediction. **b** Compensation and control process

period of T. Inevitably, this phenomenon will deteriorate the control performance of the PMSM system.

Considering the computation delay problem, Fig. 4.2 illustrates a specially-designed delay compensation strategy for the proposed numerical solution based FCS-MPCC scheme. It consists of two sequential parts: on-line delay time prediction and implementation of current pre-compensation and control.

(a) Delay prediction

In order to estimate the delay time, the proposed FCS-MPCC algorithm based on the new predicting model without delay compensation should be conducted on the PMSM at first (as in Fig. 4.2a). It can be noted that different from the traditional single sampling technique, sampling is implemented twice in each control period, one of which is still at the beginning of a period, and the other is after voltage selection but before switching state actuation. At length, the sampling currents over $t_{k-1} \sim t_k$ and $t_k \sim t_{k+1}$ are $i_{r\_1(k-1)}$, $i_{r\_2(k-1)}$ and $i_{r\_1(k)}$, $i_{r\_2(k)}$, respectively. Besides, the control voltage within τ should be the one applied in the last step. Between t_k and $t_k + \tau$, according to the d-axis predicting plant, the time delay equals the solution of (4.4):

$$i_{r\_2}(k) = \frac{m}{c} + [\frac{y}{c}\sin(p\omega_m\tau) + \frac{z}{c}\cos(p\omega_m\tau)] \cdot \exp(-\frac{R_s\tau}{L_s}) \qquad (4.4)$$

Definitely, it is tedious to solve this equation. Firstly, because τ is tiny, the variation of $\sin(p\omega_m\tau)$ is much larger than $\cos(p\omega_m\tau)$ within τ. It is appropriate to approximate the sine function to $p\omega_m\tau$ while the cosine function to 1 according to the theorem of equivalent infinitesimal replacement. Meanwhile, $\exp(-\frac{R_s\tau}{L_s})$ is equivalent to $1 - \frac{R_s\tau}{L_s}$. Then, (4.4) can be simplified as:

$$i_{r\_2}(k) = -\frac{yR_s}{cL_s} \cdot \tau^2 + (\frac{y}{c} - \frac{zR_s}{cL_s}) \cdot \tau + \frac{m+z}{c} \qquad (4.5)$$

And τ can be derived as:

$$\tau = \frac{yL_s - zR_s + \sqrt{-4L_sR_scyi_{r\_2}(k) + 4L_sR_smy + 2L_sR_szy + R_s^2z^2}}{2yR_s} \quad (4.6)$$

Practically, in order to increase the accuracy of delay estimation, Eq. (4.6) can be executed repetitiously in multiple (e.g., $N = 15$) cycles, and then, the average value would be adopted as the required result.

(b)   Implementation of current pre-compensation and control

After obtaining τ, the dual-sampling technique is unnecessarily adopted for the compensation and control process (as in Fig. 4.2b) any longer. As is illustrated in Fig. 4.3, the proposed FCS-MPCC algorithm based on the accurate plant model and compensation includes six stages at the kth instant:

(1)   State measurement: Detect the real-time phase currents, rotor position $\theta(k)$ and speed $\omega_m(k)$ and transform the three phase currents i_a, i_b and i_c to the d, q-axis currents $i_d(k)$ and $i_q(k)$ (denoting $i_{r\_1}(k)$) in Fig. 4.2b) according to $\theta(k)$.

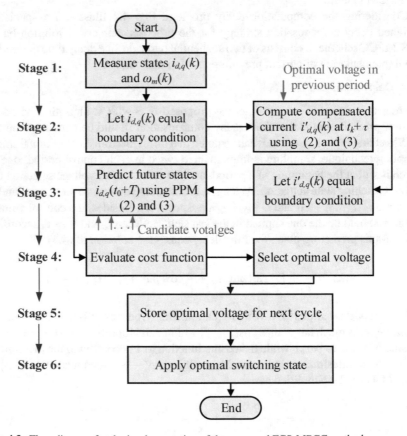

Fig. 4.3 Flow diagram for the implementation of the proposed FCS-MPCC method

(2) Pre-compensation: Let the measured currents equal the boundary condition and predict the currents $i_d(k)'$ and $i_q(k)'$ (denoting $i_r(k)'$ in Fig. 4.2b) in τ following the Eqs. (4.2) and (4.3) using the control voltage in the previous cycle, where $i'_r(k)$ represents the compensated current.

(3) Prediction: Let the compensated currents $i_d(k)'$ and $i_q(k)'$ equal the boundary condition and substitute them together with $\omega_m(k)$ into the PPM to estimate the future current states $i_d(t_0 + T)$ and $i_q(t_0 + T)$ for all the candidate manipulated voltage vectors.

(4) Evaluation: Substitute all the predicted currents one by one into a cost function and select the voltage vector that minimizes the cost function.

$$J = (i_d^* - i_d(t_0 + T))^2 + (i_q^* - i_q(t_0 + T))^2 \tag{4.7}$$

where i_d^*, i_q^* are the d, q-axis reference currents, respectively.

(5) Storage: Store the selected optimal voltage vector which will be used in stage 2 in the next control period.

(6) Switching state application: Single out the corresponding switching state according to the best voltage vector and apply it to the system.

Theoretically, the real current at $t_{k+1} + \tau$ will reach $i_{s(k+1)}$, indicating that the proposed method obeys the optimum control principle.

4.1.4 Verifications

Experiments are carried out to verify the proposed FCS-MPCC strategy on a surface-mounted PMSM drive whose parameters are consistent with those in Table 4.1. The algorithms are implemented on a DSP TMS320X28335 control board. For the sake of comprehensive analysis, the system is tested at the control frequencies of 2 kHz and 1 kHz, respectively. Finally, it needs to be mentioned that in each control period, the codes of both the novel and conventional FCS-MPCC methods are executed once.

(a) Test results at 2 kHz

Table 4.1 Motor and control parameters

Parameter	Value	Unit
stator winding resistance R_s	0.6383	Ω
stator inductance $L_s = L_d = L_q$	2	mH
the number of pole pairs p	4	–
moment of inertia J	0.013	kg·m^2
voltage constant C'_e	5.01	–
permanent magnet flux linkage Ψ_f	0.085	Wb
sampling time T	0.0001	s

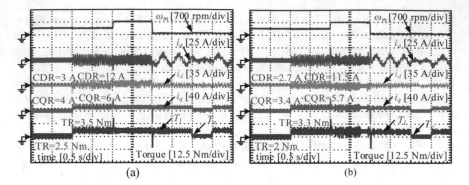

(a)   (b)

Fig. 4.4 Experimental results at control frequency of 2 kHz. **a** FCS-MPCC based on Euler discretization method. **b** FCS-MPCC based on proposed prediction plant without delay compensation

On the one hand, in order to compare the control performance of the traditional discretization method and the proposed strategy, both the two algorithms are verified firstly without using any compensation techniques. The experimental setup is as follows: the machine speeds up from standstill to 350 rpm (medium speed) between 0 and 1 s, after which it stabilizes in the next 1 s. At 1 s, the rated load is imposed on the shaft suddenly, and from 2 s, the reference speed is set as 700 rpm (high speed). At 3 s, the speed decreases to 50 rpm (ultra-low speed), after which the speed will remain at this level until 5 s. Moreover, the load is suddenly removed and then applied again at 4 s and 4.5 s respectively. In order to compare the calculation delays (complexity) of the proposed and conventional FCS-MPCC algorithms, the execution time is tested offline by using the code execution time measurement function of the processor (with an emulator). Figure 4.4 shows the experimental results of the traditional and the new methods without compensation. At first, it can be seen both algorithms show remarkable dynamics over the low and medium speed range. Specifically, the settling time (from 0 to 350 rpm) is shorter than 0.1 s and the system shows strong robustness against the external load disturbance. Whereas, when the machine is controlled to approach the rated point, although the rising time for the two methods is similar, the settling time for the new strategy is slightly longer (around 0.7 s) from the perspective of torque, and after 2.7 s, the torque ripples witness a visible decline. Then, the steady-state control performance of the two methods is different, which is mainly reflected in the d-axis current ripples (CDR), q-axis current ripples (CQR) and torque ripples (TR). Firstly, the CDR, CQR and TR of the traditional method are 3 A, 4 A and 2.5 Nm under the no-load condition at 350 rpm, respectively. While they decrease by around 0.3 A (10%), 0.6 A (15%) and 0.5 Nm (16%) when the new algorithm is applied. Moreover, the same trend can be witnessed under load conditions. The CDR, CQR and TR drop from 12 A, 6 A and 3.5 Nm for the traditional discretization method to 11.5 A, 5.7 A and 3.3 Nm for the proposed PPM. These represent that the novel method can improve the steady-state control performance in the LCF conditions. Finally, the execution delay time for the novel and traditional methods are 0.0327 ms and 0.0302 ms, respectively, proving that the computation complexity of the proposed FCS-MPCC method is slightly higher than that of the traditional one.

On the other hand, in order to verify the effectiveness of the brand-new on-line delay compensation strategy. The experimental setup is designed as follows: the machine is controlled by the new approach without compensation between 0 and 1 s, in which the delay time is estimated. Then, after 1 s, the proposed method with delay compensation is adopted for control. Firstly, Table 4.2 records the estimated calculation delay (containing sampling consumption) in fifteen control periods. It can be seen that the differences among those data are small (maximum error is 0.0018 ms), so the assumption in Sect. 4.1.2 is reasonable. Besides, the average delay time can be calculated as 0.032 ms, being very close to the off-line test value. Then, Fig. 4.5 shows that under the no-load conditions, the CDR, CQR and TR get down to 2.45 A, 3.1 A and 1.85 Nm when the calculation delay algorithm is implemented. Moreover, in comparison with Fig. 4.4b, the control ripples for the integrated method also experience a slight decrease (4.3% CDR, 3.5% CQR and 4.5% TR) when the rated load is applied. In order to verify the effectiveness of the

Table 4.2 Time delay in fifteen different periods

kth period	Delay (ms)	kth period	Delay (ms)	kth period	Delay (ms)
1	0.0312	6	0.0328	11	0.0312
2	0.0322	7	0.0322	12	0.0326
3	0.0310	8	0.0313	13	0.0322
4	0.0325	9	0.0318	14	0.0322
5	0.0312	10	0.0326	15	0.0326

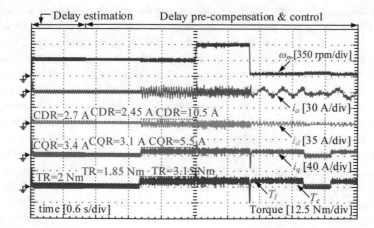

Fig. 4.5 Experimental results of the proposed strategy considering calculation delay compensation at control frequency of 2 kHz

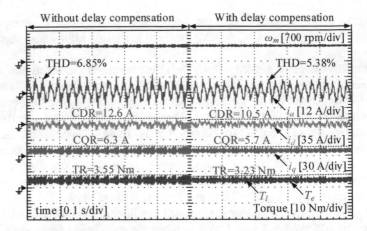

Fig. 4.6 Comparison of control performance before and after calculation delay compensation at control frequency of 2 kHz

proposed delay compensation strategy more intuitively, Fig. 4.6 compares the steady-state performance before and after delay compensation when the machine operates at the rated point. Same as the results over the medium-speed range, the CDR, CQR and TR experience a visible decline as well. Besides, the total harmonic distortion (THD) of phase current before delay compensation is 6.85%, which is about 27.3% higher than that after delay compensation. These verify that the delay compensation strategy is capable of reducing the current and torque ripples so as to enhance the performance of the proposed FCS-MPCC strategy at the control frequency of 2 kHz.

(b)   Test results at 1 kHz

Since that the calculation delay time has been illustrated previously, this part will directly compare the system performance among the traditional Euler discretization-based FCS-MPCC, the proposed numerical solution-based FCS-MPCC without delay compensation and the proposed method with delay compensation. The experimental setups are consistent with the above-mentioned ones. Figure 4.7 illustrates the comparison results of the traditional and the proposed method without delay compensation. Firstly, in comparison with the results at the control frequency of 2 kHz, the CDR, CQR and TR become much larger regardless of the working conditions. Secondly, similar to Fig. 4.4, the proposed FCS-MPCC shows better steady-state performance than the traditional approach. In detail, the CDR, CQR and TR of the traditional method are 8 A, 9.5 A and 5.5 Nm under the no-load condition at 350 rpm, respectively, while they decrease to 7 A, 8.6 A and 5.15 Nm when the new strategy is applied. And the same trend occurs for the load conditions. Figure 4.8 demonstrates the experimental results of the proposed strategy with calculation delay compensation. Under the no-load condition at 350 rpm, the CDR, CQR and TR are further smaller than those in Fig. 4.7b, with the values of 6.8 A, 8.4 A and 5.12 Nm, respectively. Compared to the results at the control frequency of 2 kHz, it can be seen that when the control frequency is lower, the delay compensation effect will

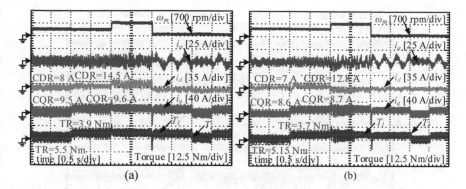

(a)   (b)

Fig. 4.7 Experimental results at control frequency of 1 kHz. **a** FCS-MPCC based on Euler discretization method. **b** FCS-MPCC based on proposed prediction plant without delay compensation

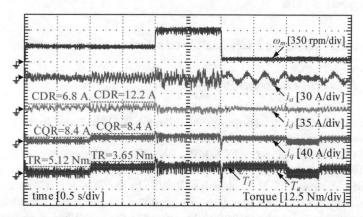

Fig. 4.8 Experimental results of the proposed strategy considering calculation delay compensation at control frequency of 1 kHz

get relatively less significant because the low control frequency contributes more to the system performance degradation. In Fig. 4.9, the steady-state performance of the proposed method without and with delay compensation at the rated point is compared. The THD of the phase current between 0 and 0.5 s is 9.25% and it drops to 8.78% between 0.5 s and 1 s. Besides, similar to the results in Fig. 4.6, the ripples of d, q-axis currents and torque get smaller after compensation as well.

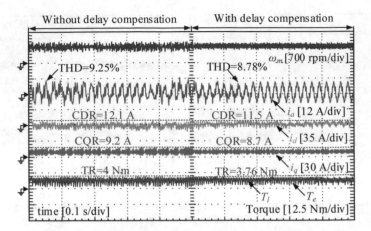

Fig. 4.9 Comparison of control performance before and after calculation delay compensation at control frequency of 1 kHz

4.2 Multi-objective FCS-MPC with Delay Compensation

4.2.1 Problem Descriptions

As illustrated in Chap. 1, concerning the different TCOs included in the cost functions, there exist two types of commonly used FCS-MPC techniques: multi-objective FCS-MPC and single-objective FCS-MPC. The FCS model predictive torque control (FCS-MPTC), whose TCOs are the torque and flux, is the typical application of the multi-objective FCS-MPC theories [9, 10]. In the FCS-MPTC-based control systems, extra speed controllers (e.g., PI controller) are required for tracking the reference speed ω_m^* and generating the reference torque. When it comes to the single-objective FCS-MPC approaches, the MPCC-based controller in which the current is the only TCO is the most intensively studied topic [11, 12]. Similar to the FCS-MPTC method, the speed regulators cannot be eliminated either because they should be used for q-axis reference current i_q^* generation. Unluckily, the speed controllers contribute much to the complexity of the drive topology, and it is an uphill task to tune the parameters of them because of lacking theoretical design procedures for the multi-type-controller based systems. In order to simplify the control topology, [13–15] employ a new single-objective MPC method by only using one FCS-MPSC-based controller. This method only contains the speed control loop in the topology, resulting in that neither the steady-state nor dynamic characteristics of the other machine states (e.g., current and torque) will be evaluated simultaneously. Besides, [16–18] employ one single multi-objective FCS-MPC controller to control both currents and speed simultaneously. To improve the control performance of this kind of method, these studies focus on developing different disturbance and noise suppression techniques, leading to the fact that the cost function is relatively complex. It also needs to be mentioned that although this kind of method has been proven to be effective to improve the speed dynamics, many crucial challenges still require further discussion. For instance, scholars never stop looking for analytical weighting factor tuning methods and novel calculation delay handling strategies to enrich the correlative theories.

A main purpose of this part is to design an FCS-MPC-based controller that can meanwhile take multiple TCOs (speed and d, q-axis currents) into account, achieving high-performance single-closed-loop control structure. Different from [16–18], the cost function of the new method is simplified without considering the external disturbances. However, the same to the traditional multi-objective FCS-MPC methods, the combination of the two variables in one single cost function is not a straightforward task because they are of different properties (units, magnitude in values and state sensitivity to voltage change). Weighting factors must be introduced into the cost function for balancing the different control targets. Unfortunately, there are no ready-made standard analytical or numerical methods or control design theories to tune these parameters, and they are mainly determined based on empirical procedures [19]. Consequently, the efficient weighting factor handling strategies should be specially developed for the novel model predictive speed and current controller.

Another typical problem of an FCS-MPC for PMSM drive systems is that loads of calculations have to be executed during the switching state selection process, so the calculation time is long [20, 21]. The time delay between the state measurement and actuation would deteriorate the system performance if not considered, lessening the concept and effect of optimal control. The TSP strategy that uses the machine model shifted one step forward to calculate the manipulated voltages is explained in [22] and [23]. Although this scheme is simple to implement, it is not totally effective to accurately determine the optimum switching state because its implementation prerequisite is that the time delay equals to a control period.

This part proposes a multi-objective FCS-MPC strategy based on novel calculation delay compensation method to achieve high-performance speed and current control. A hybrid cost function, which is based on the speed and current errors is developed to select the best voltage vector. In comparison with traditional single-objective FCS-MPC, no extra speed controllers are employed any more, simplifying the control topology greatly. Moreover, the reasons why the weighting factors have to be employed for the proposed MPC controller are explained explicitly, and an efficient handling strategy that contains two sequential implementation procedures is developed. Finally, the performance (especially steady-state performance) of the currents and output torque can be improved by using the proposed delay compensation method based on dual sampling in one control period.

4.2.2 Improved Model for Multi-objective FCS-MPC

State-space model which comprises no less than one differential equation is capable of accurately reflecting the transient behaviors of a multivariable system, making itself well-suited for PMSM MPC applications with further reference to the non-linearization and strong coupling properties. Equation (4.1) only illustrates the electrical performance of the surface-mounted PMSM, which is not the general model of the PMSM. For the sake of comprehensiveness, in this section, the general electrical and mechanical properties of the PMSMs are incorporated into the state-space model:

$$\frac{di_d}{dt} = -\frac{R_s}{L_d}i_d + \frac{L_q}{L_d}p\omega_m i_q + \frac{u_d}{L_d} \tag{4.8}$$

$$\frac{di_q}{dt} = -\frac{L_d}{L_q}p\omega_m i_d - \frac{R_s}{L_q}i_q + \frac{u_q}{L_q} - \frac{\Psi_f}{L_q}p\omega_m \tag{4.9}$$

$$\frac{d\omega_m}{dt} = \frac{1}{J}(1.5p(\Psi_f i_q + (L_d - L_q)i_d i_q) - T_l) \tag{4.10}$$

where L_d, L_q are dq-axis inductance, and $L_d = L_q = L_s$ for a surface-mounted PMSM. T_l is the load torque. J is the overall inertia considering load.

In order to predict the future states, the continuous domain model must be discretized in a time step of T (control period). When the forward Euler discretization is applied to (4.8) and (4.9), the PPM can be expressed as:

$$i_d(k+1) = \frac{L_d - TR_s}{L_d}i_d(k) + \frac{TL_q p}{L_d}\omega_m(k)i_q(k) + \frac{T}{L_d}u_d(k) \tag{4.11}$$

$$i_q(k+1) = -\frac{TL_d p}{L_q}\omega_m(k)i_d(k) + \frac{L_q - TR_s}{L_q}i_q(k)$$
$$+ \frac{T}{L_q}u_q(k) - \frac{T\Psi_f p}{L_q}\omega_m(k) \tag{4.12}$$

where $i_d(k)$, $i_q(k)$ and $\omega_m(k)$ are the states at the kth sampling instant. $i_d(k+1)$ and $i_q(k+1)$ are the predicting values at the $(k+1)$th period (t_{k+1}). The discretization operation cannot be applied to (4.10) because the speed alteration within the kth period is not directly determined by the manipulated variables $u_d(k)$ and $u_q(k)$, but it is a direct consequence of the overall currents in the cycle rather than just the initial currents at t_k. Considering this, take definite integral of (4.10) between t_k and t_{k+1} and the PPM for speed estimation turns:

$$\omega_m(k+1) = \frac{1.5p\Psi_f}{J}\int_{t_k}^{t_{k+1}} i_q dt + \frac{L_d - L_q}{J}\int_{t_k}^{t_{k+1}} i_d i_q dt$$
$$- \frac{1}{J}\int_{t_k}^{t_{k+1}} T_l dt + \omega_m(k) \tag{4.13}$$

where $\omega_m(k+1)$ is the speed at t_{k+1}. Usually, the control period T is very short, so an appropriate equivalence that i_q and $i_d i_q$ experience linear changes under the control of $u_d(k)$ and $u_q(k)$, and the external load torque T_l remains constant at $T_l(k)$ over the period can be made. In this case, (4.13) can be rewritten as:

$$\omega_m(k+1) = \frac{3p\Psi_f T}{4J}(i_q(k+1) - i_q(k)) - \frac{T_l(k)T}{J} + \omega_m(k)$$
$$+ \frac{(L_d - L_q)T}{2J}(i_d i_q(k+1) - i_d i_q(k)) \tag{4.14}$$

Obviously, when predicting the speed state, the future currents should be calculated by (4.11) and (4.12) in advance. It should be addressed that only by doing this can $\omega_m(k+1)$ be associated with the manipulated voltages.

4.2.3 Implementation of Multi-objective FCS-MPC

The implementation procedures of an FCS-MPC algorithm can be summarized as follows: the measured currents and speed as well as the seven manipulated voltages are substituted into the PPM to predict the next step's states, and then the predicted values are evaluated by a cost function so as to select the optimal control voltage to be applied. In this process, the cost function serves as the key component for optimizing calculation, and it determines whether an FCS-MPC-based controller is a so-called single-objective or multi-objective scheme. In order to clearly compare the differences between a single-objective controller and the proposed approach, the cost function of an FCS-MPCC-based controller, which only regards the current as the control target, is demonstrated.

$$J_{SO} = (i_d^* - i_d(k+1))^2 + (i_q^* - i_q(k+1))^2 \tag{4.15}$$

4.2.3.1 Structure of Multi-objective FCS-MPC Method

Figure 4.10 depicts the structure of a dual objective (speed and current) FSC-MPC strategy. In comparison with the single-objective FCS-MPCC method, the explanations of the new algorithm are detailed as follows.

Most crucially, the speed and current performance is assessed by a novel multi-objective cost function:

$$J_{MO} = k_1(i_d^* - i_d(k+1))^2 + k_1(i_q(k-1) - i_q(k+1))^2 + k_2(\omega_m^* - \omega_m(k+1))^2 \tag{4.16}$$

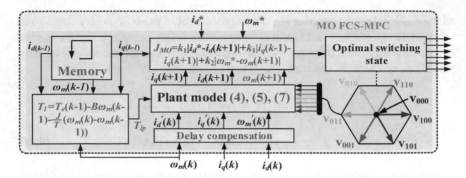

Fig. 4.10 Structure of multi-objective FCS-MPC-based controller

It can be noted that the proposed algorithm aims to track the d-axis current and speed references (i_d^* and ω_m^*), but an extra term of absolute error between the previous q-axis current $i_q(k-1)$ and $i_q(k+1)$ is also incorporated into the cost function. The reason for this configuration includes that it is impossible to simultaneously and precisely compel the rotating speed and q-axis current to track the manually set references in a PMSM by one sole MPC-based controller, and hence $(i_q^*-i_q(k+1))^2$ cannot be integrated into the cost function. But from the other point of view, when the machine reaches the equilibrium state in the control process, the q-axis current ripples are expected to be furthest reduced so as to improve the steady-state performance of the PMSM. Practically, this performance characteristic can be evaluated by the current variations in the adjacent intervals. For the purpose of lowering the fluctuations of the q-axis current which is directly related to the output electromagnetic torque, we import $(i_q(k-1)-i_q(k+1))^2$ into the multi-objective cost function. It should be mentioned that because the cost function is able to evaluate both the current and speed characteristics, it is unnecessary to track a q-axis reference, so the speed controller is eliminated in the multi-objective FCS-MPC control structure.

Secondly, a load torque observer is integrated into the proposed controller. In Eq. (4.14), the real-time load torque should be used for predicting the future speed information. But a physical load torque sensor is usually not installed in the PMSM for the sake of cost reduction. In this case, a new issue that how to determine $T_l(k)$ arises. Wang et al. [24] presents a theoretical solution to the problem by designing an observer to estimate T_l, but it does not explain the explicit discretization procedures. In this part, considering that the sampling period T is short (typically 0.2 ms), we make an appropriate assumption that the contemporary load torque is equal to that in the $(k-1)$th cycle. Then, a discrete torque estimator can be deduced by applying backward Euler technique to the mechanical dynamic equation, that is:

$$T_{lp}(k) = T_e(k-1) - B\omega_m(k-1) - \frac{J}{T}(\omega_m(k) - \omega_m(k-1)) \qquad (4.17)$$

where T_{lp} is the observed load torque, and T_e is the output electromagnetic torque of the machine,

$$T_e(k-1) = 1.5p(\Psi_f i_q(k-1) + (L_d - L_q)i_d(k-1)i_q(k-1)) \qquad (4.18)$$

Thirdly, the time delay caused by substituting the seven alternative voltages one by one into the plant model to calculate the future states is compensated, which will be beneficial to the system performance. The detailed compensation strategy is explained in Sect. 4.2.3.3.

So far, the system control structure becomes simple by using a single multi-objective FCS-MPC-based controller to simultaneously control speed and currents. However, the weighting factors k_1 and k_2 must be adopted to balance the differences between the two kinds of control targets. To solve this problem, Sect. 4.2.3.2 proposes a brand-new handling method based on SVN and balance of state sensitivity to voltage alteration.

4.2.3.2 Weighting Factor Handling Strategy

One important reason why the weighting factors k_1 and k_2 are essential for the proposed multi-objective FCS-MPC-based controller is that the nature (unit and magnitude in value) of the speed and d, q-axis currents is different. Besides, the magnitude of the speed and current variations corresponding to the same voltage alteration is different. In other words, the sensitivity of the current and speed (response variables) to the voltage variations (control variables) is different. Unlike an FCS-MPTC method of which TCOs (torque and flux) are totally independent, resulting in that the importance of the two TCOs to switching state selection varies, the TCOs (speed and currents) of the proposed strategy are corelated. Specifically, the speed alteration is the consequence of current variation, as is illustrated by the machine model (4.10). Therefore, the importance of the speed and currents in the cost function should be the same without considering their sensitivity to voltage variations. Aiming at these two aspects, the weighting factor tuning strategy for the proposed multi-objective FCS-MPC-based controller can be achieved by following two sequential steps: (a) SVN, and (b) Balance of state sensitivity to voltage alteration.

(a)   SVN

The definition of SVN can be described as follows: the state variables, including current, speed and voltage, are expressed as the fractions of the defined base quantities. After SVN, the large imparity in the magnitude and units of different variables can be removed. Namely, the representations of all variables in the system become uniform with per unit values (the unit is pu and the magnitude of values ranges from -1 to 1). In this part, the base quantities for PMSM system normalization are specially designed as follows. Firstly, the base voltage U_0 is defined as the maximum line-to-line voltage:

$$U_0 = U_{dc} \tag{4.19}$$

Then, according to this normalization principle, the base current I_0 is:

$$I_0 = \frac{2P_{rated}}{\sqrt{3}U_{dc}} \tag{4.20}$$

where P_{rated} is the rated power of the motor. The base speed ω_{m0} equals the value of rotor angular velocity (in rad/s) at which the machine develops 1 pu voltage at its terminals with zero current. According to the relationship between the back electromotive force (EMF) and the rotor speed, the base speed can be derived as:

$$\omega_{m0} = \frac{U_{dc}}{\sqrt{3}k_0 C_e' \Psi_f} \tag{4.21}$$

where C_e' is the voltage constant, and k_0 is an adjusting constant.

After applying SVN to the drive system, the normalized state variables are:

$$i_{dn} = \frac{i_d}{I_0}, i_{qn} = \frac{i_q}{I_0}, u_{dn} = \frac{u_d}{U_0}, u_{qn} = \frac{u_q}{U_0}, \omega_{mn} = \frac{\omega_m}{\omega_{m0}} \tag{4.22}$$

Then the electrical and mechanical PPMs can be expressed as:

$$\mathbf{x}_{en}(k+1) = \mathbf{A}_e \mathbf{x}_{en}(k) + \mathbf{B}_e \mathbf{u}_n(k) + \mathbf{d}_{en}(k) \tag{4.23}$$

$$\mathbf{x}_{mn}(k+1) = \mathbf{A}_m \mathbf{x}_{mn}(k) + \mathbf{A}_{ms}(\mathbf{x}_{en}(k+1) - \mathbf{x}_{en}(k)) + \mathbf{d}_{mn}(k) \tag{4.24}$$

where $\mathbf{A}_e = \begin{bmatrix} \frac{L_d - TR_s}{L_d} & 0 \\ 0 & \frac{L_q - TR_s}{L_q} \end{bmatrix}, \mathbf{B}_e = \begin{bmatrix} \frac{T}{L_d} & 0 \\ 0 & \frac{T}{L_q} \end{bmatrix}, \mathbf{A}_m = \begin{bmatrix} 0 & 0 \\ 0 & 1 \end{bmatrix}$ and $\mathbf{A}_{ms} = \begin{bmatrix} 0 & 0 \\ 0 & \frac{3p\Psi_f T}{4J} \end{bmatrix}$. $\mathbf{x}_{en} = [i_{dn}, i_{qn}]^T, \mathbf{u}_{en} = [u_{dn}, u_{qn}]^T, \mathbf{x}_{mn} = [0, \omega_{mn}]^T$, and $\mathbf{d}_{en}$ and $\mathbf{d}_{mn}$ are considered as the disturbances which can be written as:

$$\mathbf{d}_{en} = [\frac{TL_q p}{L_d} \omega_{mn} i_{qn}, -\frac{TL_d p}{L_q} \omega_{mn} i_{dn} - \frac{T\Psi_f p}{L_q} \omega_m]^T \tag{4.25}$$

$$\mathbf{d}_{mn} = [0, \frac{(L_d - L_q)T}{2J}(i_d i_q(k+1) - i_d i_q(k))]^T \tag{4.26}$$

(b)   Balance of state sensitivity to voltage alteration

For the sake of simplicity, assume that $L_d = L_q = L_s$ and ignore the effects of mutual inductance. The machine model (4.12) and (4.14) can be expressed as:

$$i_q(k+1) = \frac{L_s - TR_s}{L_s}i_q(k) - \frac{T\Psi_f p}{L_s}\omega_m(k) + \frac{T}{L_s}u_q(k) \qquad (4.27)$$

$$\omega_m(k+1) = \frac{3p\Psi_f T}{4J}(i_q(k+1) - i_q(k)) - \frac{T_l T}{J} + \omega_m(k) \qquad (4.28)$$

In addition to $u_q(k)$, a voltage disturbance $\Delta u_q(k)$ is assumed to be applied to the machine at t_k. It can be deduced that the current alteration $\Delta i_q(k+1)$ caused by $\Delta u_q(k)$ in one cycle can be described as:

$$\Delta i_q(k+1) = \frac{T}{L_s}\Delta u_q(k) \qquad (4.29)$$

While the speed alteration $\Delta\omega_m(k+1)$ caused by $\Delta u_q(k)$ is:

$$\Delta\omega_m(k+1) = \frac{3p\Psi_f T}{4J}\Delta i_q(k+1) \qquad (4.30)$$

Obviously, the currents in the machine are more sensitive to the voltage changes. In order to balance the state sensitivity to voltage variation differences between the two variables, the weighting factors can be set as:

$$k_1 = 1, \quad k_2 = \frac{4J}{3p\Psi_f T} \qquad (4.31)$$

Then, the magnitude of the speed and current variations caused by the same voltage change within one control period will stand at the same level. Note that the weighting factor k_2 is related to the sampling period T, representing that the discretization scheme is an important factor that influences the optimization operation of the multi-objective FCS-MPC-based controller.

4.2.3.3   Novel Computation Delay Compensation

The goal of the cost function is to suppress the current and speed tracking errors to the most degree, but it will be blocked by the calculation delay. For the sake of convenience, this part takes i_d as an example to analyze the issue, and a comparison between the ideal condition and the real condition is shown in Fig. 4.11. It can be seen that between t_{k-1} and t_k, $\mathbf{v}_{010}$ is evaluated as the best voltage vector with the measured information available at the beginning of this period. However, the actuation action is not applied at t_{k-1} but at $t_{k-1} + \tau$, where τ is the computation delay, leading to that the current locus cannot be controlled as expected and some deviation occurs after a control period. In this case, $\mathbf{v}_{010}$ may not be the best solution, and this phenomenon cannot be self-healed in the later cycles, such as the $(k+1)$th and $(k+2)$th period.

The traditional TSP strategy is proposed with an assumption of that $\tau = T$ and the concrete compensation procedures can be referred to in [23]. But in practice, the delay

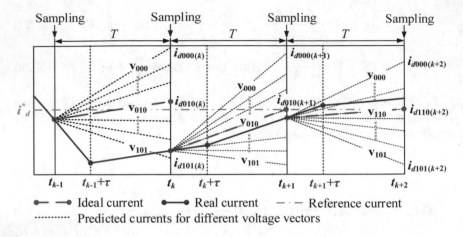

Fig. 4.11 Comparison between the ideal condition and the real condition

time cannot exceed a control period, as shown in Fig. 4.11. In view of the problem, this section presents a compensation scheme on the basis of dual sampling and delay time estimation (as in Fig. 4.12a). Another remarkable feature can be noticed in Fig. 4.12b that instead of operating on the predicted states, the new method will calibrate the measured values before predicting the future states.

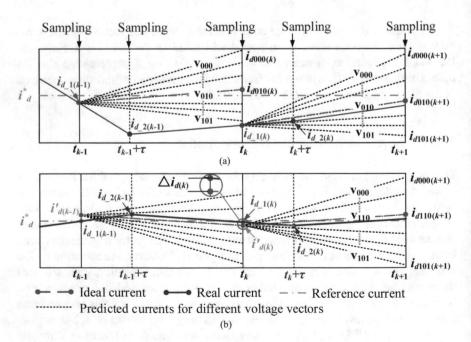

Fig. 4.12 Proposed compensation method. **a** Delay time calculation. **b** Compensation

(a) τ estimation

In order to estimate and compensate the delay time, assume that the real current will shift linearly in accordance with the predicted trend and the computation delay remains constant in each control period.

When estimating τ, the proposed multi-objective FCS-MPC algorithm without compensation will be applied. But an obvious feature can be witnessed that there are two sampling points in each control period, one of which is at the start of a period. The other is at the end of voltage selection, where the computation work is finished. In Fig. 4.12a, the sampling currents over $t_{k-1} \sim t_k$ and $t_k \sim t_{k+1}$ are $i_{d\_1(k-1)}$, $i_{d\_2(k-1)}$ and $i_{d\_1(k)}$, $i_{d\_2(k)}$, respectively. The delay can be calculated by:

$$\tau = \frac{|i_{d\_2(k)} - i_{d\_1(k)}|}{|i_{d\_2(k)} - i_{d\_2(k-1)}|} \cdot T \qquad (4.32)$$

(1)   Compensation

Since τ is obtained, the detailed compensation procedures are as follows at t_k:

(a)   First measurement and *abc/dq* transformation: use sensors to detect the real phase currents, rotor position and speed and transform the measured phase currents to the *dq*-axis currents.

(b)   Compensation: predict the current variation $\Delta i_{d(k)}$ in τ (see Fig. 4.12b).

$$\Delta i_{d(k)} = \frac{i_{d\_1(k)} - i_{d\_2(k-1)}}{T - \tau} \cdot \tau \qquad (4.33)$$

$i_{d\_2(k-1)}$ is also the sampling current in the last interval, and the current used for prediction is:

$$i'_{d(k)} = i_{d\_1(k)} + \Delta i_{d(k)} \qquad (4.34)$$

Following the above approach, the compensated q-axis current $i_q{}'(k)$ and speed $\omega_m{}'(k)$ can also be calculated. Then, the control process is:

(c)   Prediction: use $i_d{}'(k)$, $i_q{}'(k)$, $\omega_m{}'(k)$ and T_l to estimate the future states for all the seven candidate voltage vectors.

(d)   Evaluation: substitute all the predicted values into the cost function and determine the optimal voltage vector.

(e)   Second measurement: repeat step (a) to get currents ($i_{d\_2}(k)$, $i_{q\_2}(k)$), position and speed $\omega_{m\_2}(k)$.

(f)   Switching state application: apply the corresponding optimum switching state to the drive.

Theoretically, the real current at $t_k + \tau$ will reach the previously predicted value, which means that the proposed multi-objective FCS-MPC method obeys the optimum control rule just with a delay of τ.

4.2.4   Verifications

The performance of the proposed multi-objective FCS-MPC without and with delay compensation is tested and compared by the means of simulation and experiments. The motor and control parameters of the PMSM prototype are consistent with those in Table 4.1.

4.2.4.1   Simulation Results

For the sake of comprehensive discussion, the following verifications are included in this section. Firstly, in order to comparatively explain the control performance, apart from the proposed multi-objective FCS-MPC method, the traditional double closed-loop FCS-MPCC strategy (see Fig. 4.13) is tested as well. Secondly, to validate the effectiveness of the proposed calculation delay technique, the performance characteristics of the proposed method without and with compensation are compared. Thirdly, to verify that the proposed calculation delay strategy is superior to the traditional TSP technology, the steady-state current performance of the proposed FCS-MPC method compensated by the two methods is analyzed.

   The verification procedures for the traditional and proposed methods without delay compensation are as follows: the machine speeds up from standstill to 100 rad/s at first, after which it will stabilize until 1.0 s when a constant load of 5 Nm is suddenly applied. From 1.5 s, the reference speed is set as 200 rad/s, while the motor begins to slow down from 2.5 s. After removing the external load at 3.0 s, the rotating speed of the machine continues to decline until it reaches zero.

(a)   Analysis on the state-state performance

Figures 4.14 and 4.15 illustrates the control performance of the traditional and proposed multi-objective FCS-MPC algorithms without delay compensation. For the two methods, it can be noted that the motor speed remains stable after it arrives

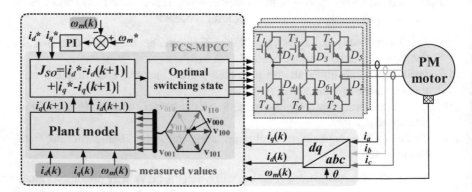

Fig. 4.13 Structure of an FCS-MPCC-based drive system

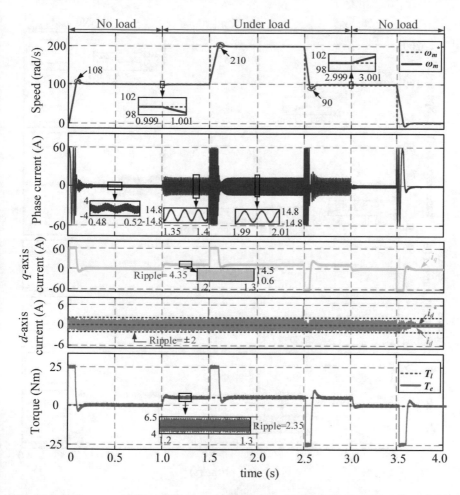

Fig. 4.14 Performance of the traditional double closed-loop FCS-MPCC algorithm without delay compensation

at the setpoint no matter whether the machine is loaded or not during the whole test range. Besides, the average value of the d-axis current can level off at zero. These indicate that the proposed multi-objective FCS-MPC has marked speed and current tracking capability without steady-state errors. However, visible current and torque ripples can be witnessed in both Figs. 4.14 and 4.15. Specifically, under load conditions, the peak-to-peak ripple of q-axis for the traditional method is 4.35 A, equaling nearly 33.8% of the average value (12.55 A), and the d-axis current and torque ripples are about 4 A and 2.35 Nm, respectively. Comparatively speaking, the d, q-axis current and torque ripples of the proposed multi-objective method are 4 A, 4.5 A (37.5% of the average value) and 2.5 Nm, respectively. As for the no-load cases, the magnitude of the current and torque ripples stands at very similar

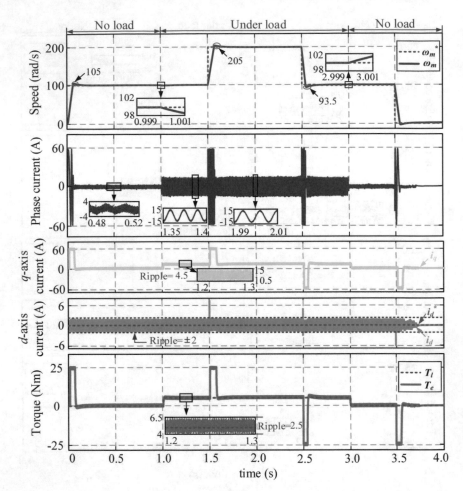

Fig. 4.15 Performance of the proposed multi-objective FCS-MPC algorithm without delay compensation

position with rated load conditions for both of the methods. It can be noted that the proposed multi-objective FCS-MPC controller has as good steady-state performance as the traditional double closed-loop FCS-MPCC method, indicating that the proposed multi-objective FCS-MPC-based controller is totally qualified for regulating speed as well as current without using a speed controller to generate q-axis reference current.

(b)   Analysis on the dynamic performance

Dynamic performance evaluation of the control system needs to consider the following aspects, namely, settling time, speed and current overshoot and robustness to abrupt load variation.

During acceleration, firstly, the rise time of the traditional FCS-MPCC method is nearly 0.15 s while it is about 0.1 s for the proposed MPC algorithm regardless of load or no-load conditions. Secondly, between 0 and 0.5 s, the speed overshoot in Figs. 4.14 and 4.15 is 8 (8%) and 5 rad/s (5%), respectively, and between 1.5 and 2 s, it is 10 and 5 rad/s, respectively. These represent that the proposed method has slightly better dynamics than the traditional one, complying with the results in [16]. Moreover, because the output of the speed controller (q-axis reference) of the traditional method is constrained as 60 A, both the phase and q-axis currents just jump to the safety level during acceleration for both of methods. When the q-axis current reaches 60 A, the maximum output electromagnetic torque is about 25 Nm. Moreover, the d-axis current fluctuates clearly at 1.5 s, which is caused by cross-coupling effect. In terms of deceleration, the settling time and speed overshoot are about 0.15 s and 10 rad/s (10%) at 2.5 s for the traditional method and for the multi-objective strategy, they are 0.1 s and 6.5 rad/s (6.5%), respectively. In this process, the largest phase and q-axis current are both nearly −60 A, which stands at the opposite position with acceleration. Similarly, the d-axis current witnesses some fluctuations at 2.5 and 3.5 s. When the load is imposed and removed on the motor, only a speed deviation of 2 rad/s appears for both the traditional method and the proposed strategy, and their output electromagnetic torques can quickly reaches the expected level. These illustrate that the proposed multi-objective FCS-MPC algorithm has as strong robustness to torque variations as the double closed-loop structure. Overall, from the perspective of dynamics, the proposed multi-objective method is qualified for high-performance control.

Figure 4.16 demonstrates the simulation results of the proposed FCS-MPC algorithm with delay compensation. As to the experimental setup, the algorithm is implemented to calculate the delay τ before 0.25 s, after which the proposed algorithm launches. Compare to Fig. 4.15, the advantage of the improved method is mainly reflected by the steady-state performance (the dynamic performance is similar), especially the current and torque ripples. In detail, before 0.25 s, the ripples of q-axis current, d-axis current and output torque ripples are 6 A, 4 A and 2 Nm, respectively. Whereas, they become nearly 5 A, 3.56 A and 1.75 Nm with a drop of 16.7, 11 and 12.5% after compensation. The phase current ripple also experiences a visible declination before and after 0.25 s. Overall, the compensation method is conducive to high-performance control drive.

In order to more intuitively observe the control performance of the proposed method, Table 4.3 gives the THD simulation results for the phase currents under the rated load condition. At the speed of 100 rad/s, the phase current THD for the traditional double closed loop FCS-MPCC method without compensation, proposed multi-objective FCS-MPC method without compensation and proposed method with compensation is 3.53%, 3.85% and 2.26%, respectively, and at the speed of 200 rad/s, they are 5.83%, 5.93% and 4.35%, respectively. In the first place, the performance of the proposed FCS-MPC method is very similar to the traditional strategy, further proving that the proposed method is of high performance. Secondly, after delay compensation, the current performance is greatly improved, indicating that the calculation delay is a key factor that influences the steady-state performance. Finally, it is

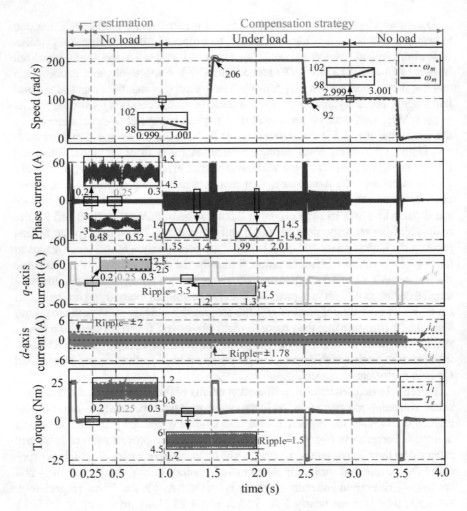

Fig. 4.16 Performance of the proposed multi-objective FCS-MPC algorithm with delay compensation

Table 4.3 Phase current THD under load condition

Method Type	Working States (rad/s)	THD (%)
Double loop FCS-MPCC without compensation	100	3.52
	200	5.83
Multi-objective FCS-MPC without compensation	100	3.85
	200	5.93
Multi-objective FCS-MPC with compensation	100	2.26
	200	4.35

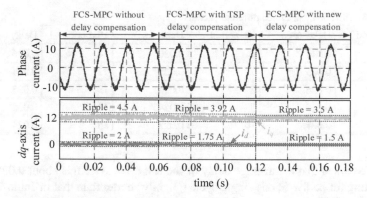

Fig. 4.17 Current performance under the control of proposed multi-objective FCS-MPC with different delay compensation methods

interesting to see that the THD at the rated operation point is larger than that at the low speed cases. Many reasons have caused this phenomenon, one of which is the dead-time effect.

To compare the performance of the traditional TSP delay compensation method and the novel one, the following verification setup is designed: the machine rotates at the speed of 100 rad/s with rated load. The proposed multi-objective FCS-MPC algorithm without delay compensation is used before 0.06 s, in which the time delay is computed. Between 0.06 and 0.12 s, the algorithm with TSP compensation (as in 22]) is adopted, and afterwards, the proposed algorithm is employed to control the motor. Figure 4.17 illustrates the current performance in the whole control process. It can be seen that both of the compensation methods are effective to suppress the current ripples so as to improve the steady-state performance. But it deserves to be mentioned that the new compensation method is superior to the traditional one because it obeys the optimal control principle. Specifically, the d, q-axis current ripples are 1.75 A and 3.92 A for the TSP compensation strategy while they are 1.5 A and 3.5 A respectively for the new technique. In term of the phase current THD, it is 2.85% for the conventional compensation method, which is slightly larger than that (as in Table 4.3) for the novel dual-sampling based strategy.

4.2.4.2 Experimental Results

Apart from comparing the performance of the traditional FCS-MPCC method and the proposed multi-objective FCS-MPC method, the experiment will also be carried out to compare the control performance before and after delay compensation. For the sake of analytical simplicity, the delay time will be tested using the proposed delay estimation strategy beforehand in this part. Under no load condition, the delay time is measured when the machine is controlled by the multi-objective FCS-MPC algorithm. Table 4.4 records the delay value in fifteen different control periods. It

Table 4.4 Time delay in fifteen different periods

kth period	Delay (ms)	kth period	Delay (ms)	kth period	Delay (ms)
1	0.024	6	0.026	11	0.026
2	0.025	7	0.025	12	0.024
3	0.026	8	0.024	13	0.025
4	0.025	9	0.024	14	0.025
5	0.027	10	0.023	15	0.024

should be noted that the average delay of the test drive system is about 0.0258 ms, accounting for 25.8% of one single cycle (slightly shorter than that in Table 4.2 due to simplicity). Undoubtedly, this will influence the control performance.

By contrast to the simulation procedures, the experimental setup without considering online delay estimation is as follows: the machine speed is set as 100 rad/s under no-load condition at first, after which it will maintain constant until 4.0 s when the rated load of 5 Nm is applied. From 6.0 s, the machine speed rises to 200 rad/s, while the motor decelerates at 10.0 s. At 12.0 s, the load is removed, and then the reference speed is set as 10 rad/s (ultra-low speed) at 14 s. Figures 4.18, 4.19 and 4.20 show the experimental results of the traditional FCS-MPCC method, the proposed multi-objective FCS-MPC method without delay compensation and the proposed FCS-MPC approach with delay compensation, respectively.

On the one hand, similar to the simulation results, Figs. 4.18 and 4.19 illustrate that the proposed method has as good steady-state performance as the FCS-MPCC method regardless of the machine speed and load conditions. Firstly, the machine speed can track the targeted values only with small fluctuations (less than 2 rad/s) in the stable states and the average value of the d-axis current is zero ($i_d^* = 0$). No static errors are seen, so the speed and current tracking properties for the two methods are good. Secondly, in terms of the q-axis current, they are both slightly higher than zero under the no-load conditions, which is caused by the friction, they level off at nearly 12 A with small fluctuations under load cases. Thirdly, the output electromagnetic torque of the machine in Figs. 4.18 and 4.19 is consistent with the external load, indicating that the PMSM drive system can work normally under the rated load conditions. Finally, when it comes to the variable ripples, the d, q-axis current and torque ripples of the proposed method are 9 A, 11 A and 3.5 Nm, and for the FCS-MPCC strategy, they are 8.6 A, 10.3 A and 3.2 Nm, respectively. In accord with the simulation results, the steady-state performance differences between the two methods are small. On the other hand, interestingly, the settling time of the traditional FCS-MPCC method and the proposed strategy is very similar for the test bench. Over the low-speed range (0–100 rad/s), it is about 0.1 s while it is nearly 0.3 s over the higher speed range (100–200 rad/s). This happens because a relatively smaller q-axis current has been generated during acceleration from 100 to 200 rad/s in the experiment. But the deceleration characteristics show a pretty similar trend to the simulation results. Moreover, due to the extra speed controller used in the topology of FCS-MPCC method, the speed experiences small overshoot regardless

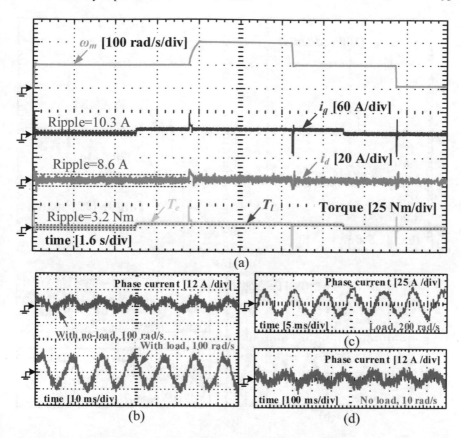

Fig. 4.18 Experimental performance of the traditional double closed-loop FCS-MPCC algorithm without delay compensation. **a** Speed, *d*-axis current, *q*-axis current and torque. **b** Phase currents under no-load and load conditions at 100 rad/s. **c** Phase current under load condition at 200 rad/s. **d** Phase current under no-load condition at 10 rad/s

of acceleration and deceleration in Fig. 4.18, but it is modest in Fig. 4.19. In terms of the system robustness against load disturbances, the machine speed can remain stable even if the load is suddenly imposed or removed for the new method, which is the same to the double loop approach. Finally, when the *q*-axis current reaches ± 60 A in the transient process, the output electromagnetic torque in Figs. 4.18 and 4.19 increases to the same level (about ± 24.5 Nm) and the *d*-axis current shows similar changes as well. Overall, the dynamics of the new method is remarkable as well.

By comparing Figs. 4.19 and 4.20, slight reductions in the current and torque harmonics and ripples can be witnessed for the algorithm with delay compensation. Taking the condition that the machine runs at 100 rad/s as an example, the *q*-axis, *d*-axis and torque ripples of the approach without delay compensation under the no-load conditions are about 11 A, 9 A and 3.5 Nm, respectively, but they are 9.8 A, 8.3 A and 3.2 Nm in Fig. 4.20, with a decrease of 10.9%, 7.8% and 8.6%, respectively.

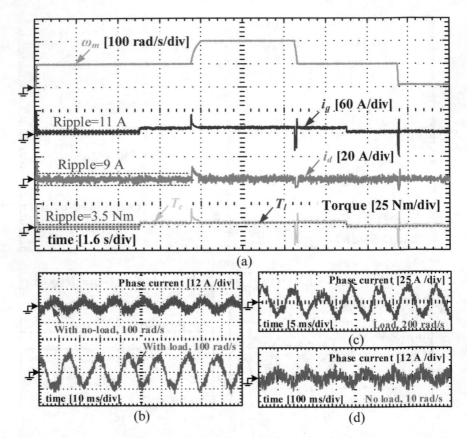

Fig. 4.19 Experimental performance of the proposed multi-objective FCS-MPC algorithm without delay compensation. **a** Speed, d-axis current, q-axis current and torque. **b** Phase currents under no-load and load conditions at 100 rad/s. **c** Phase current under load condition at 200 rad/s. **d** Phase current under no-load condition at 10 rad/s

Meanwhile, the results under load conditions see the similar trend to those under the no-load conditions. As far as the phase currents (Fig. 4.19b–d and Fig. 4.20b–d) are concerned, the harmonics have been reduced visibly after the proposed delay compensation method is applied. These indicate that the proposed delay compensation method is effective for ripple reduction and performance improvement. In order to clearly observe the phase current ripples, Table 4.5 shows the experimental THD analysis results under the rated load condition. Firstly, compared to the values in Table 4.4, the magnitude of THD is larger. The phenomenon arises mainly because the back EMF of the PMSM used for test is more like a square wave rather than a sine wave, and this should be also responsible for the low sinusoidal degree of phase current. It is expected that the proposed method is more effective for a PMSM with standard sine-wave back EMF. Secondly, similar to the simulation results, the THD

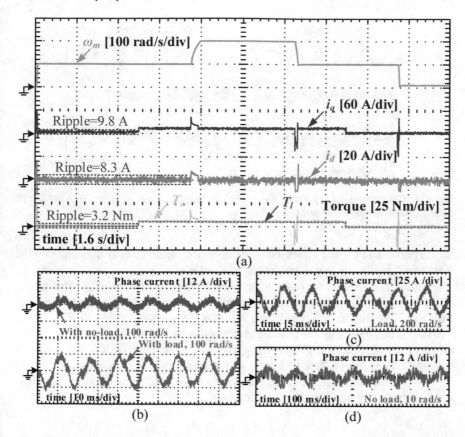

Fig. 4.20 Experimental performance of the proposed multi-objective FCS-MPC algorithm with delay compensation. **a** Speed, d-axis current, q-axis current and torque. **b** Phase currents under no-load and load conditions at 100 rad/s. **c** Phase current under load condition at 200 rad/s. **d** Phase current under no-load condition at 10 rad/s

Table 4.5 Phase current THD under load condition

Method type	Working States (rad/s)	THD (%)
Double loop FCS-MPCC without compensation	100	8.52
	200	11.26
Multi-objective FCS-MPC without compensation	100	9.45
	200	12.1
Multi-objective FCS-MPC with compensation	100	7.96
	200	10.63

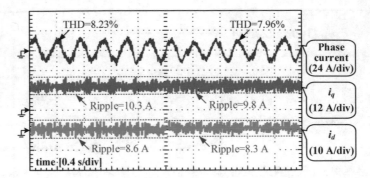

Fig. 4.21 Current performance under the control of proposed multi-objective FCS-MPC with different delay compensation methods

for the proposed FCS-MPC method with delay compensation is minimal among the three cases, indicating that the proposed method is effective.

Figure 4.21 illustrates the experimental results (speed is 100 rad/s and load torque is 5 Nm) when the proposed multi-objective FCS-MPC is compensated by the traditional TSP-based (between 0 and 2 s) and the proposed dual-sampling-based (between 2 and 4 s) delay compensation strategies. Similar to the simulation results, the new method shows a slightly better performance. In detail, the THD of the phase current declines from 8.23% to 7.96%, and d, q-axis current ripples drop by nearly 3.5% and 4.9%, respectively. These indicate that the novel compensation strategy is totally suitable for the FCS-MPC applications.

4.3 Summary

This Chapter presents effective techniques to improve the prediction accuracy of the FCS-MPC methods. In detail, firstly, a numerical-solution-based PPM is proposed to improve the control performance of the FCS-MPC method used in the surface-mounted PMSMs in the LCF cases. Secondly, a multi-objective (speed and current) FCS-MPC strategy is proposed for the PMSMs, achieving simple single-closed-loop control topology. To improve the accuracy of the proposed multi-objective FCS-MPC controller, a novel computation delay compensation method is developed to lower the ripples and harmonics in the system. The main contributions of this chapter can be summarized as:

(1) An accurate machine PPM based on numerical solutions is proposed to eliminate the diverse side effects caused by the traditional linear Euler discretization algorithm. By using the novel model, the precise future states can be calculated, ensuring that the optimal control voltage can be selected precisely. Besides, instead of using the traditional calculation delay compensation method, a specially designed compensation technique based on delay estimation and

current pre-compensation is investigated for the numerical-solution-based FCS-MPC method.

(2) Compared to the traditional single-objective FCS-MPC algorithms, both speed and current are included into the cost function of the proposed FCS-MPC controller as the TCOs. By doing this, the speed controller is no longer needed. When designing the algorithms, an improved PMSM model which contains not only differential but also integral equations is established, and an effective weighting factor handling strategy on account of SVN and balance of state sensitivity to voltage alteration is developed. Moreover, in order to suppress the current and torque ripples caused by the computation delay, a compensation scheme on the basis of dual sampling is proposed. The novel delay compensation scheme is composed of two sequential procedures: delay time estimation and compensation. The experimental results illustrate that the state ripples decrease by around 8% after compensation and the proposed delay compensation strategy has better performance in comparison to the traditional TSP technique.

References

1. Z.Q. Zhu, Y. Liu, Analysis of Air-gap field modulation and magnetic gearing effect in fractional-slot concentrated-winding permanent-magnet synchronous machines. IEEE Trans. Ind. Electron. **65**(5), 3688–3698 (2018)
2. J. Lu, X. Zhang, Y. Hu, J. Liu, C. Gan, Z. Wang, Independent phase current reconstruction strategy for ipmsm sensorless control without using null switching states. IEEE Trans. Ind. Electron. **65**(6), 4492–4502 (2018)
3. C. Gong, Y. Hu, G. Chen, H. Wen, Z. Wang, K. Ni, A DC-bus capacitor discharge strategy for PMSM drive system with large inertia and small system safe current in EVs. IEEE Trans. Ind. Inform. **15**(8), 4709–4718 (Aug. 2019)
4. X. Zhang, L. Zhang, Y. Zhang, Model predictive current control for pmsm drives with parameter robustness improvement. IEEE Trans. Power Electron. **34**(2), 1645–1657 (2019)
5. J. Liu, C. Gong, Z. Han, H. Yu, IPMSM model predictive control in flux-weakening operation using an improved algorithm. IEEE Trans. Ind. Electron. **65**(12), 9378–9387 (2018)
6. Y. Zhang, B. Xia, H. Yang, Performance evaluation of an improved model predictive control with field oriented control as a benchmark. IET Electr. Power App. **11**(5), 677–687 (2017)
7. S. Zhao, X. Huang, Y. Fang, J. Zhang, Compensation of DC-Link voltage fluctuation for railway traction PMSM in multiple low-switching-frequency synchronous space vector modulation modes. IEEE Trans. Veh. Technol. **67**(1), 235–250 (2018)
8. Z. Wang, B. Wu, D. Xu, N.R. Zargari, A current-source-converter-based high-power high-speed PMSM drive with 420-Hz switching frequency. IEEE Trans. Ind. Electron. **59**(7), 2970–2981 (2012)
9. Y. Wang et al., Deadbeat model-predictive torque control with discrete space-vector modulation for PMSM drives. IEEE Trans. Ind. Electron. **64**(5), 3537–3547 (2017)
10. P. Cortes, J. Rodriguez, C. Silva, A. Flores, delay compensation in model predictive current control of a three-phase inverter. IEEE Trans. Ind. Electron. **59**(2), 1323–1325 (2012)
11. Y. Yang, H. Wen, D. Li, A fast and fixed switching frequency model predictive control with delay compensation for three-phase inverters. IEEE Access **5**, 17904–17913 (2017)

12. X. Xiao, Y. Zhang, J. Wang, H. Du, An improved model predictive control scheme for the PWM rectifier-inverter system based on power-balancing mechanism. IEEE Trans. Ind. Electron. **63**(8), 5197–5208 (2016)

13. W. Xie et al., Finite-control-set model predictive torque control with a deadbeat solution for PMSM drives. IEEE Trans. Industr. Electron. **62**(9), 5402–5410 (2015)

14. F. Ban, G. Lian, J. Zhang, B. Chen, G. Gu, Study on a novel predictive torque control strategy based on the finite control set for PMSM. IEEE Trans. Appl. Superconductivity **29**(2), 1–6 (March 2019). (Art no. 3601206)

15. Y. Yan, S. Wang, C. Xia, H. Wang, T. Shi, Hybrid control set-model predictive control for field-oriented control of VSI-PMSM. IEEE Trans. Energy Convers. **31**(4), 1622–1633 (2016)

16. W. Tu, G. Luo, R. Zhang, Z. Chen, R. Kennel, "Finite-control-set model predictive current control for PMSM using grey prediction," *2016 IEEE Energy Conversion Congress and Exposition (ECCE)*, (Milwaukee, WI, 2016), pp. 1–7

17. M. Preindl, S. Bolognani, Model predictive direct speed control with finite control set of PMSM drive systems. IEEE Trans. Power Electron. **28**(2), 1007–1015 (2013)

18. M. Preindl, S. Bolognani, "Model predictive direct speed control with finite control set of PMSM-VSI drive systems," *2011 Workshop on Predictive Control of Electrical Drives and Power Electronics*, (Munich, 2011), pp. 17–23

19. C. Gong, Y. Hu, K. Ni, J. Liu, J. Gao, SM load torque observer based FCS-MPDSC with single prediction horizon for high dynamics of surface-mounted PMSM. IEEE Trans. Power Electron. **35**(1), 20–24 (2020)

20. A. Formentini, A. Trentin, M. Marchesoni, P. Zanchetta, P. Wheeler, Speed finite control set model predictive control of a PMSM fed by matrix converter. IEEE Trans. Industr. Electron. **62**(11), 6786–6796 (2015)

21. E.J. Fuentes, C.A. Silva, J.I. Yuz, Predictive speed control of a two-mass system driven by a permanent magnet synchronous motor. IEEE Trans. Industr. Electron. **59**(7), 2840–2848 (2012)

22. F. Mwasilu, H.T. Nguyen, H.H. Choi, J. Jung, Finite set model predictive control of interior pm synchronous motor drives with an external disturbance rejection technique. IEEE/ASME Trans. Mechatron. **22**(2), 762–773 (2017)

23. P. Cortes et al., "Guidelines for weighting factors design in model predictive control of power converters and drives," *2009 IEEE International Conference on Industrial Technology*, Gippsland, VIC, 2009, pp. 1–7

24. Y. Wang et al., Deadbeat model-predictive torque control with discrete space-vector modulation for PMSM drives. IEEE Trans. Industr. Electron. **64**(5), 3537–3547 (2017)

Chapter 5
MPC Accuracy Improvement for PMSMs—Part II

Yaofei Han, Chao Gong⑩, and Jinqiu Gao

This chapter can also be divided into two parts, which focuses on improving the accuracy of the model predictive control (MPC) method.

In Chap. 4, to solve the accuracy problem of the traditional predicting plant model (PPM), a numerical-solution-based finite control set model predictive current control (FCS-MPCC) method is developed. However, this method is only suitable for the surface-mounted permanent magnet synchronous motors (PMSMs) because the numerical solutions to the general machine model (4.8) and (4.9) are too complicated to implement. Hence, in order to improve the prediction accuracy of a general PMSM or interior PMSM (IPMSM), a sub-step FCS-MPCC strategy is developed in Sect. 5.1. Besides, considering that the flux linkage might change in practice, degrading the accuracy of the PPM. This Chapter develops a direct handling strategy by using the parameter identification technology. In detail, a stable sliding mode flux linkage (SM-FL) observer is designed to detect the real-time flux linkage. It can be noticed that the solution to the flux linkage mismatch issue is similar to those in Chap. 3. This means that the parameter-identification strategies are able to improve not only the robustness but also prediction accuracy.

As is shown in Chap. 1, the linearization problem also influences the accuracy of the MPC method. In order to solve the problem, a brand-new linearization approach is proposed to handle the strong coupled nonlinear IPMSM mathematical model. In detail, an improved linear plant model (ILPM) which is suitable for the motor with constant load torque is established. It deserves to be mentioned that in Sect. 5.2, the proposed ILPM is adopted to achieve a continuous control set MPC (CCS-MPC) strategy, which is used in the IPMSM flux-weakening operations.

Y. Han
National Maglev Transportation Engineering R&D Center, Tongji University, Shanghai 201804, China

C. Gong (✉) · J. Gao
School of Automation, Northwestern Polytechnical University, Xi'an 710072, China

© The Author(s), under exclusive license to Springer Nature Singapore Pte Ltd. 2022
Y. Han et al. (eds.), *Model Predictive Control for AC Motors*,
https://doi.org/10.1007/978-981-16-8066-3_5

5.1 Flux-Observer-Based Sub-Step FCS-MPCC

5.1.1 Problem Description

Being similar to the background illustrated in Sect. 4.1.1, in practice, in order to reduce the current and torque ripples of the IPMSMs, high control frequency (HCF) needs to be employed when implementing an FCS-MPCC algorithm. But it should be acknowledged that the high switching frequency will generate large amount of switching loss, decreasing the system efficiency. On this ground the relatively low control frequency (LCF) drive systems are also very common in the industrial applications [1, 2]. However, there is a crucial problem for the FCS-MPCC strategies in these systems. That is, the traditional PPM obtained by using the forward Euler discretization algorithm is not accurate in the LCF cases. This will degrade the control performance undoubtedly.

Since FCS-MPCC is a model-based control method, the control performance characteristics are highly dependent on the model parameters including the flux linkage [3–5]. Once flux linkage mismatch (FLM) occurs, the steady-state errors and current/torque ripples will sharply increase undesirably. Unfortunately, the flux linkage is subject to the armature reaction and operating temperature [6], so it varies constantly. Hence, the pre-set parameter cannot not always match the real-time status. Paper [3] analyses the prediction errors caused by the parameter mismatches theoretically and uses an incremental prediction model to reduce the influences of the flux linkage. In [4], apart from an incremental model, an extended state observer (ESO) is proposed to enhance the robustness against the stator inductance mismatch. Song and Zhou [5] proposes an MPC controller design method by widening the bandwidth of the whole system, which contributes to the robustness improvement. These methods are achieved by reconstructing the prediction plant model, which requires amount of mathematical derivations and can be defined as the indirect solution to the FLM problem. Comparatively speaking, there also exists a direct handling strategy by using the parameter identification technology. Yet there are few studies focusing on this natural solution in the IPMSM MPC applications, impeding the further progress of the correlative technologies.

One main purpose of this section is to propose a novel FCS-MPCC method with tripartite calculation frequency but single actuation frequency to reduce the impacts caused by long control time. Firstly, the reasons why the traditional FCS-MPCC implementation method affects the control performance are investigated. Secondly, one control period T is divided into three equal parts, and IPMSM model is discretized in a step of $T/3$. Then, the predicting operation will be serially implemented three times for each candidate control voltage. After evaluating the predicted values by the use of a cost function and singling out the best control voltage, the control signals will be generated immediately and applied to the drive system. Additionally, for the purpose of directly coping with the FLM problem, this section designs a signum-function (SF) based sliding mode (SM) observer to detect the parameter in real time and improve the prediction accuracy. Meanwhile, a Lyapunov function is constructed to analyze the stability of the SM observer so as to obtain the explicit condition that makes the system stable.

5.1.2 Impacts of LCF and FLM

This part uses a mathematical analysis method to explain why the LCF and FLM problems influence the prediction accuracy of the FCS-MPCC method, which can be regarded as expanded descriptions of Sect. 4.1.1. Because the IPMSM is the research object in this part, the machine model used for analysis is consistent with (4.8)–(4.12).

(a) LCF influence

The prediction process (only take the d-axis current i_d as an example for simplicity) of an FCS-MPCC algorithm can be depicted in Fig. 5.1a when a candidate voltage vector is applied, where $i_d{}^*$, i_{dr} and i_{dp} are the reference, real and predicted d-axis currents, respectively. The right terms of machine model (4.8) is denoted as:

$$f(t) = -\frac{R_s}{L_d}i_d(t) + \frac{L_q}{L_d}p\omega_m(t)i_q(t) + \frac{u_d(t)}{L_d} \tag{5.1}$$

where L_d, L_q are dq-axis inductance. ω_m is the rotor mechanical angular speed. u_d is the d-axis control voltage. R_s is the stator winding resistance. p represents the number of pole pairs.

Between t_k and t_{k+1}, the physical descriptions of the traditional FCS-MPCC implementation procedures based on the first-order difference discretization method (4.11) and (4.12) are as follows: Firstly, the measured state values at t_k together with the candidate control voltages are substituted into (5.1) to calculate the slope of the tangent line of the current curve (sl), that is, $sl = f(t_k)$. Secondly, $f(t_k)$ is regarded as the slope of a straight line:

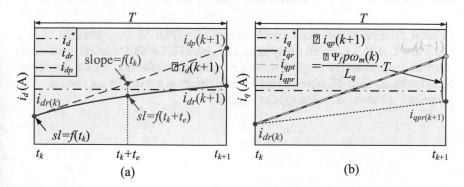

Fig. 5.1 Impacts of LCF and FLM on performance. **a** LCF influence. **b** FLM influence

$$y = k \cdot x + y_0 \tag{5.2}$$

That is, $k = f(t_k)$. Then, $i_d(k + 1)$ is calculated taking $i_d(k)$ as the benchmark within one control period. Namely, $y_0 = i_d(k)$, $x = T$. Obviously, the currents are assumed to shift linearly during T. Whereas, the effectiveness of this approximation method dilutes as T extends. In detail, because the currents in the machine alters constantly, the slope of the tangent line of the current curve cannot remain at $f(t_k)$, so the real current is nonlinear. For instance, at $t_k + t_e$, sl is supposed to equal $f(t_k + t_e)$ which is calculated by using the states at that moment rather than $f(t_k)$. Undoubtedly, the longer T is, the larger the prediction error $\triangle i_d(k + 1)$ in Fig. 5.1a gets.

(b)   FLM influence

The IPMSM model illustrates that FLM can only reflect in the q-axis current. The prediction process of the q-axis current is demonstrated in Fig. 5.1b, where i_q and i_{qr} are the reference and real q-axis currents, respectively. i_{qpt} and i_{qpr} are the predicted currents without and with parameter mismatch, respectively. When the FLM is $\triangle \Psi_f$, the error between the predicted currents without and with parameter mismatch $\triangle i_{qp}(k + 1)$ is:

$$\triangle i_{qp}(k + 1) = \frac{\triangle \Psi_f p \omega_m(k)}{L_q} \cdot T \tag{5.3}$$

where Ψ_f represents the flux linkage.

It can be noted that the FLM influence is also related to the machine speed, and the higher the speed is, the more significant the impact of FLM becomes.

Overall, considering the LCF and FLM problems, the selected control voltage might not be the optimal one in each control period, decreasing the machine control performance.

5.1.3   Flux-Observer-Based Sub-Step FCS-MPCC Strategy

In order to improve the system performance of a IPMSM drive working at LCF and with FLM, this part will firstly discuss a novel FCS-MPCC algorithm based on tripartite predictions but single actuation to reduce the influence of LCF. As for the FLM problem, a SM-FL observer is specially designed to provide the instantaneous parameter information for the predicting plant.

5.1.3.1 Implementation of the Novel FCS-MPCC Method

According to the above analysis, the predicting error caused by LCF mainly arises from the one-step linear approximation process. Besides, referring to that the estimation accuracy is still high when T is extremely short, the overall error is inclined to decline if each control period is further segmented and the prediction algorithm is implemented in each segment with the previously updated state values. On this ground a three-segment way is developed, as is shown in Fig. 5.2. At first, the IPMSM model requires to be discretized in $T/3$, and the plant model (including mechanical dynamics) for prediction based on (4.8)–(4.10) is:

$$i_d\left(k + \frac{1}{3}\right) = \frac{3L_d - TR_s}{3L_d} i_d(k) + \frac{TL_q p}{3L_d}\omega_m(k)i_q(k) + \frac{T}{3L_d}u_d(k) \tag{5.4}$$

$$i_q\left(k + \frac{1}{3}\right) = -\frac{TL_d p}{3L_q}\omega_m(k)i_d(k) + \frac{3L_q - TR_s}{3L_q} i_q(k)$$
$$+ \frac{T}{3L_q}u_q(k) - \frac{T\Psi_f p}{3L_q}\omega_m(k) \tag{5.5}$$

$$\omega_m\left(k + \frac{1}{3}\right) = \frac{Tp\Psi_f}{2J} i_q(k) + \frac{Tp(L_d - L_q)}{2J} i_d(k)i_q(k)$$
$$+ \frac{BT + 3J}{3J}\omega_m(k) - \frac{TT_l}{3J} \tag{5.6}$$

where J is the overall viscous considering load, and B is the friction coefficient.

Then, the implementation procedures of the proposed FCS-MPCC strategy between t_k and t_{k+1} are as follows:

(1) State measurement: detect the real-time d, q-axis currents $i_{dr}(k)$ and $i_{qr}(k)$, and the rotor speed $\omega_{mr}(k)$ at t_k.

(2) Current and speed prediction: firstly, substitute $i_{dr}(k)$, $i_{qr}(k)$ and $\omega_{mr}(k)$ into the predicting plant to estimate the d, q-axis currents $i_{dp}{}'(k + 1/3)$, $i_{qp}{}'(k + 1/3)$ and speed $\omega_{mp}{}'(k + 1/3)$ at $t_k + T/3$ for all possible voltage vectors. Secondly, use $i_{dp}{}'(k + 1/3)$, $i_{qp}{}'(k + 1/3)$, $\omega_{mp}{}'(k + 1/3)$ and the corresponding control

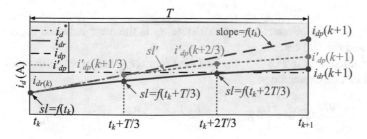

Fig. 5.2 Implementation of the proposed FCS-MPCC algorithm

voltage to estimate the future states $i'_{dp}(k + 2/3)$, $i_{qp}'(k + 2/3)$ and $\omega_{mp}'(k + 2/3)$ at $t_k + 2\ T/3$. Finally, $i_{dp}'(k + 1)$ and $i_{qp}'(k + 1)$ at t_{k+1} are calculated according to the predicted state values in the last step.

(3)   Evaluation: substitute the predicted currents into the cost function (5.7) to compute g for the current deviations.

$$J = \left| i_d^* - i'_{dp}(k + 1) \right| + \left| i_q^* - i'_{qp}(k + 1) \right| \tag{5.7}$$

(4)   Voltage and switching state selection: select the voltage vector that minimizes the cost function and determine the corresponding switching state.
(5)   Actuation: apply the optimal switching state to the system.

It can be noticed that although triple predictions occur within a control period, the actuation takes place only once, representing that the proposed method does not result in higher switching loss. Moreover, in comparison with the traditional strategy, because the slope of each approximate linear segment is closer to the slope of the tangent line of the current curve at any moment (as in Fig. 5.2), the estimation error between $i_{dp}'(k + 1)$ and $i_{dr}(k + 1)$ is much smaller than that between $i_{dp}(k + 1)$ and $i_{dr}(k + 1)$.

5.1.3.2   Proposed SM-FL Observer

SM control relies on the variable structure control technology, restraining the state variables of a system on the sliding surface [7]. Because SM control has strong robustness against the external disturbance, it has been broadly employed in the parameter identification cases. However, few researches have used SM theory to observe the flux linkage in IPMSM drives.

(a)   SM-FL observer design

According to the machine model, the flux linkage is only included in (4.9), according to the theory of SM variable structure, the q-axis current sliding surface S is needed:

$$S = \overline{i_q} = i_{q\_est} - i_q = 0 \tag{5.8}$$

where i_{q_est} is the intermediate current state. $\overline{i_q}$ is the error between the estimated current and the real current.

As is illustrated in Fig. 5.3, the SM observer for flux linkage estimation of the IPMSM can be expressed as:

$$\frac{di_{q\_est}}{dt} = -\frac{L_d}{L_q} p\omega_m i_d - \frac{R_s}{L_q} i_{q\_est} + \frac{u_q}{L_q} - \frac{p\omega_m}{L_q} \cdot kF(\overline{i_q}) \tag{5.9}$$

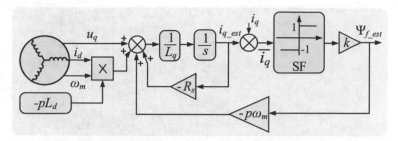

Fig. 5.3 Proposed SM-FL observer

where k is the observer gain. $F(\overline{i_q})$ is the SF-based switching function,

$$F(\overline{i_q}) = \text{sign}(\overline{i_q}) \tag{5.10}$$

When the SM observer gets to the stable state, the real-time flux linkage used for FCS-MPCC equals Ψ_{f_est}:

$$\Psi_{f\_est} = kF(\overline{i_q}) \tag{5.11}$$

(b)   Stability analysis

In order to analyze the stability condition of the proposed SM-FL observer, a function based on Lyapunov theorem is constructed as follows:

$$V = \frac{1}{2} \cdot S^T \cdot S = \frac{1}{2}\overline{i_q}^2 > 0 \tag{5.12}$$

Then, in the light of Lyapunov stability criterion, only when $\frac{dV}{dt} < 0$ will we conclude that the SM observer is stable. Take the time derivative of Eq. (5.12):

$$\frac{dV}{dt} = S^T \cdot \frac{dS}{dt} = \overline{i_q}\frac{d\overline{i_q}}{dt} \tag{5.13}$$

It can be derived according to (4.9) and (5.9) that:

$$\frac{d\overline{i_q}}{dt} = -\frac{R_s}{L_q}\overline{i_q} - \frac{p\omega_m}{L_q} \cdot (kF(\overline{i_q}) - \Psi_f) \tag{5.14}$$

Substitute (5.14) into (5.13), it can be further deduced:

$$\frac{dV}{dt} = \underbrace{-\frac{R_s}{L_q}\overline{i_q}^2}_{term1} \underbrace{- \frac{p\omega_m(kF(\overline{i_q}) - \Psi_f)}{L_q}\overline{i_q}}_{term2} \tag{5.15}$$

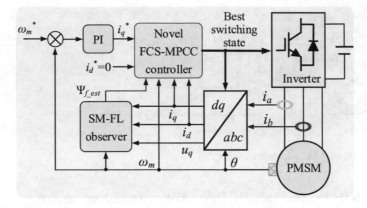

Fig. 5.4 Block diagram of the proposed control topology

Clearly, since *term*1 is less than 0, in order to make the observer stable, *term*2 should be larger than 0, that is,

$$\begin{cases} \frac{p\omega_m(k-\Psi_f)}{L_q}\overline{i_q} > 0, \ if \overline{i_q} > 0 \\ \frac{p\omega_m(-k-\Psi_f)}{L_q}\overline{i_q} > 0, \ if \overline{i_q} < 0 \end{cases} \tag{5.16}$$

Further,

$$\begin{cases} k - \Psi_f > 0, \ if \overline{i_q} > 0 \\ -k - \Psi_f < 0, \ if \overline{i_q} < 0 \end{cases} \rightarrow k > \Psi_f \tag{5.17}$$

Therefore, if k is set as a positive value that is larger than the maximum flux linkage, the SM observer will keep stable. In this section, k is five times of the flux linkage generated by the permanent magnet (PM) installed in the machine.

Overall, the novel SM-FL observer-based IPMSM FCS-MPCC topology is depicted in Fig. 5.4. In this section, a proportional integral (PI) speed controller is employed to generate the q-axis reference and the d-axis reference current is set as zero.

5.1.4 Verifications

The performance of the proposed FCS-MPCC and SM-FL observer is tested by simulation. The system parameters of the IIPMSM prototype are listed in Table 5.1.

Firstly, in order to verify the effectiveness of the proposed FCS-MPCC method, the experimental setup is as follows: firstly, the machine is controlled to speed up from standstill to 1500 rpm (medium speed) under no load condition, after which the speed will stabilize at that level until 1.5 s. At 1.0 s, a sudden load of 5 Nm

Table 5.1 Motor and control parameters

Parameter	Value	Unit
Stator winding resistance R_s	0.055	Ω
d-axis inductance L_d	0.3	mH
q-axis inductance L_q	0.8	mH
The number of pole pairs p	3	-
Moment of inertia J	0.0036	kg·m^2
Viscous coefficient B	0.0035	-
Permanent magnet flux linkage Ψ_f	0.0876	Wb
Voltage constant C'_e	3.14	–
DC-bus voltage U_{dc}	310	V

is imposed on the shaft. Then, the machine continues to accelerate from 1.5 s and the speed reference is 3000 rpm (high speed). Finally, the speed reference is set as 500 rpm (low speed) from 2.0 s, and in this process, the deceleration performance characteristics are tested. Figure 5.5 shows the control performance of the proposed method. It can be noticed that the proposed MPCC controller is able to track the reference speed as well as currents regardless of the load or no-load conditions, indicating the new algorithm can ensure good steady-state performance. The SM-FL observer can accurately observe the flux linkage in the machine, but an interesting phenomenon can be seen that the estimated flux linkage is closely related to the machine speed. As the rotor speed goes, the flux linkage ripples decrease visibly. As far as the system dynamic performance is concerned, firstly, the rise time is short, with about 0.25 s when the rotor speed rises from zero to 1500 rpm, and meanwhile, the speed overshoot is about 50 rpm (3%). When the machine decelerates, although a visible overshoot of 70 rpm (14%) is experienced, the machine returns stable after

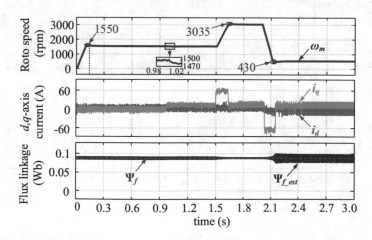

Fig. 5.5 Control performance of the proposed FCS-MPCC algorithms

about 0.8 s. Finally, the PMSM drive has strong robustness to the load variations because the speed changes little at 1.0 s and it recovers soon (nearly 0.3 s).

Secondly, this part compares the differences between the proposed method and the tradition one. Between 0 and 3 s, the machine speed is controlled by the proposed FCS-MPCC controller and the traditional controller to operate at 1500 rpm, respectively. At 1.5 s, the 5 Nm load is applied to the motor. In Fig. 5.6a, b, the proposed algorithm shows lower current (take the q-axis current as an example) ripples (QCR) and torque ripples (TR). Specifically, the CQR and TR are $\pm$ 10 A and $\pm$ 4.5 Nm for

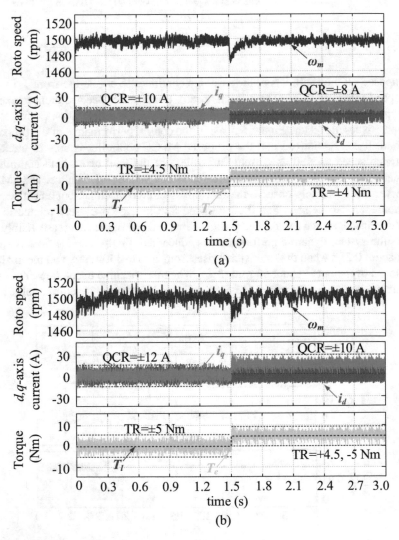

Fig. 5.6 Comparison results of the proposed method and the traditional method. **a** Proposed method. **b** Traditional method

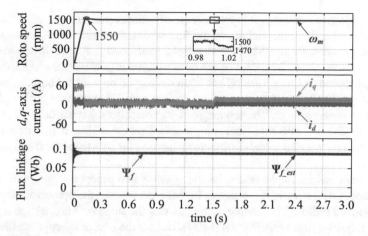

Fig. 5.7 Control performance of the proposed FCS-MPCC algorithms with 10% lower flux linkage

the proposed algorithm under the no-load condition, respectively, while they increase to ± 12 A and ± 5 Nm for the conventional method. Similar trend can be witnessed under the load conditions.

Finally, in order to further verify the SM-FL observer, assume that the machine flux linkage drops 10% during the control process. The machine is controlled by the new method to grow to 1500 rpm (as in Fig. 5.7). The SM observer is capable of accurately detecting the machine flux linkage, representing that it can be used to handle the FLM problem in practice.

5.2 Linearized CCS-MPC for IPMSM Flux-Weakening Control

5.2.1 Problem Description

In general, the conventional flux-weakening strategy based on controlling d-axis stator current is required in many drive systems for the high-speed operation [8, 9]. That method uses several proportional integrate (PI) controllers in the cascaded control loops to calculate dq-axis current and voltage. Although it is relatively simple to bring into effect, three inevitable shortcomings exist [10–13]. Firstly, because the outputs' rise speed of PI controllers cannot be constrained, the phase current might soar unexpectedly during the starting process, resulting in potential risks. The second problem is that those (six, at least) parameters of PI controllers have not gotten specific physical interpretation. Therefore, complicated parameter tuning schemes for PI controllers are usually needed, not only enlarging time cost in practical applications, but also burdening the overall control performance optimization for various objectives. Thirdly, it is experimentally discovered that the phase currents fluctuate obviously in the vicinity of the flux-weakening basic speed turning point, particularly when an IPMSM is under light load condition, which has not been significantly improved by means of the conventional flux-weakening algorithm.

As illustrated in Chap. 1, MPC has been developed on a model basis as an alternative algorithm to the PI control which does not utilize the model [14], making it probable to overcome the shortcomings of PI controllers. Mynar et al. [15] incorporates CCS-MPC into flux-weakening control schemes. But it is a direct speed control method which demands further investigation for both theoretical and practical progress.

MPC can be best implemented for the systems that accept a representation by a linear model with constraints, because, in that case, most of the optimization process can be moved offline [17], leading to the fact that the standard MPC design methods usually require a linear plant model (LPM). But for the lack of appropriate linearization approaches, LPMs for the strong coupled nonlinear IPMSMs have not notably progressed yet, limiting the control performance to some extent.

The constraint handling capability of MPC is that it can deal with the input, output and state constraints numerically over the finite horizon. While designing a MPC controller, two constraint subcategories are supposed to be considered. The best-known is to artificially set the maximum and minimum values of variables, protecting devices from overload. Limitation on the change rate of manipulated variables is directly related to the steady-state and dynamic performance likewise [18], but it is often overlooked in the previous studies.

In this part, a modulator is retained and all PI controllers are replaced with a single linear multiple-input multiple-output (MIMO) predictive controller [15], whereas the reference signals conclude not only rotor angular speed but also d-axis current. Control of d-axis current contributes to achieving the goal of adjusting accurately air-gap field magnitude to different speed settings. The other crucial aims of this part are to propose a linearization approach with high accuracy to obtain an ILPM which is expressed in state-space equations, and to introduce new constraint pattern about manipulated variable change rate.

5.2.2 Classic MPC-Based Flux-Weakening Algorithm

As is illustrated in Sect. 4.2, to use one single MPC controller to drive a motor, the PPM should include both the electrical and mechanical properties. For the sake of simplicity, the PPM in this part is directly obtained by using forward Euler discretization method. Hence, in addition to (4.11) and (4.12), the speed prediction equation should be incorporated [19]:

$$\omega_m(k+1) = \frac{1.5Tp\Psi_f}{J}i_q(k) + \frac{1.5Tp(L_d - L_q)}{J}i_d(k)i_q(k)$$

$$+ \frac{BT + J}{J}\omega_m(k) - \frac{TT_l}{J} \tag{5.18}$$

The classic IPMSM MPC algorithm is explained in terms of three different main techniques: linearization, control topology and constraints. As said in the previous section, standard MPC design requires a LPM which can be expressed in the form

$$\mathbf{x}(k+1) = \mathbf{Ax}(k) + \mathbf{Bu}(k)$$
$$\mathbf{y}(k) = \mathbf{Cx}(k) + \mathbf{Du}(k) \tag{5.19}$$

where $\mathbf{x}$ is the vector of n state variables, $\mathbf{u}$ represents the manipulated variables, $\mathbf{y}$ is a vector of the plant outputs. $\mathbf{A}, \mathbf{B}, \mathbf{C}$ and $\mathbf{D}$ are the plant coefficient matrices. However, the nonlinear terms $\omega_m(k)i_q(k)$, $\omega_m(k)i_d(k)$ and $i_d(k)i_q(k)$ become the biggest obstacle to convert (4.11) and (4.12) and (5.18) into (5.19). There are two classical approaches to remove the nonlinearities.

The first way is achieved by choosing proper state variables. It needs to neglect the difference between L_d and L_q, in other words, assume $L_d = L_q$, so that the mechanical Eq. (5.18) is linear because $i_d(k)i_q(k)$ is equal to zero. Meanwhile, those nonlinear terms in (4.11) and (4.12) can be considered as measured disturbances because $\omega_m(k)$, $i_d(k)$ and $i_q(k)$ could be measured at each sampling period [16]. Now, a LPM can be obtained when the state variables of IPMSM are the currents i_d, i_q and the speed ω_m, along with $\overline{\omega_m i_q}$, $\overline{\omega_m i_d}$:

$$\mathbf{x}(k) = [i_d(k), i_q(k), \omega_m(k), \overline{\omega_m i_d}(k), \overline{\omega_m i_q}(k)]^T \tag{5.20}$$

Aside from selecting $\overline{\omega_m i_q}$ and $\overline{\omega_m i_d}$ as the state variables, the second traditional linearization way is to assume a constant angular speed for the nonlinear terms in the whole prediction horizon [15], that is $\omega_m(k) = \Omega_{\text{mi}}$, and L_d equals to L_q as well. The electrical (4.11) and (4.12) can be described by

$$i_d(k+1) = \frac{L_d - T R_s}{L_d}i_d(k) + \frac{T L_q p \Omega_{\text{mi}}}{L_d}i_q(k) + \frac{T}{L_d}u_d(k) \tag{5.21}$$

$$i_q(k+1) = -\frac{T L_d p \Omega_{\text{mi}}}{L_q}i_d(k) + \frac{L_q - T R_s}{L_q}i_q(k) + \frac{T}{L_q}u_q(k)$$
$$-\frac{T \Psi_f p}{L_q}\omega_m(k) \tag{5.22}$$

The state variables are only i_d, i_q and ω_m:

$$\mathbf{x}(k) = [i_d(k), i_q(k), \omega_m(k)]^T \tag{5.23}$$

Although the IPMSM model is now linear, there still exists difficulties in obtaining a LPM on account of T_l. The previous study has introduced a torque observer and treated it as an unmeasured disturbance calculated by the equivalent q-axis current i_{ql} [20, 21],

$$T_l = 1.5 p \Psi_f i_{q1}(k) \tag{5.24}$$

Certainly, i_{q1} must be added to the state variable list. For example, if the second linearization method is employed, (5.23) will turn into $\mathbf{x}(k) = [i_d(k), i_q(k), \omega_m(k), i_{q1}(k)]^T$. To be honest, that method is not one hundred percent effective as a consequence of two points, one of which is that the torque of IPMSM, unlike surface PMSM, is not only relevant to q-axis current but d-axis current. The approximate observer might be far from adapting to the flux-weakening cases. Secondly, the value of i_{q1} is unknown. Only when another permanent magnet synchronous generator whose currents are measured works as the load machine could T_l be estimated precisely, but the common method is to let i_{ql} equals to i_q and this is only an approximation. Aiming to a constant load torque, an improved algorithm is proposed in Sect. 5.2.3.

As far as the classic control topology demonstrated in [15] is concerned, it can be found that an inverter controlled by space vector pulse width modulation (SVPWM) algorithm is employed to reduce the current and torque ripples, but rotating speed is the single reference signal during the natural MPC-based field-weakening operation. Comparatively speaking, conventional PI-based flux-weakening strategy is realized by controlling the d-axis current at the same time. In that course, the negative reference current i_{dref} is maintained by utilizing closed-loop cascade method [8]. Fewer references in classic algorithm undoubtedly simplify the system structure and take full advantage of the cost functions of MPC controllers. But for lack of penalty on i_d, the optimization procedure might generate unstable control voltages, issuing in output torque ripple and speed fluctuation in the constant power zone of IPMSM.

Finally, controlling an IPMSM always needs to refuse both overcurrent and overvoltage, so that the famous current and voltage limit equations are welcomed in flux-weakening operation.

$$\sqrt{u_d^2 + u_q^2} \le U_{\max} \tag{5.25}$$

$$\sqrt{i_d^2 + i_q^2} \le I_{\max} \tag{5.26}$$

where U_{max} and I_{max} are the maximum allowable voltage and current, respectively. Unfortunately, (5.24) and (5.25) cannot be modeled as a function of constraints straightly because they are nonlinear. But thanks to SVPWM technique, the voltage and current constraints can be expressed in hexagon shape

$$\begin{bmatrix} -1 & -1 & 0 & 0 & 1 & 1 \\ \frac{-1}{\sqrt{3}} & \frac{1}{\sqrt{3}} & \frac{2}{\sqrt{3}} & \frac{-2}{\sqrt{3}} & \frac{-1}{\sqrt{3}} & \frac{1}{\sqrt{3}} \end{bmatrix}^{\mathrm{T}} \begin{bmatrix} u_d(k) \\ u_q(k) \end{bmatrix} \le U_{\max} \times \mathbf{R} \tag{5.27}$$

$$\begin{bmatrix} -1 & -1 & 0 & 0 & 1 & 1 \\ \frac{-1}{\sqrt{3}} & \frac{1}{\sqrt{3}} & \frac{2}{\sqrt{3}} & \frac{-2}{\sqrt{3}} & \frac{-1}{\sqrt{3}} & \frac{1}{\sqrt{3}} \end{bmatrix}^{\mathrm{T}} \begin{bmatrix} i_d(k) \\ i_q(k) \end{bmatrix} \leq I_{\max} \times \mathbf{R} \tag{5.28}$$

where $\mathbf{R} = [1\ 1\ 1\ 1\ 1\ 1]^{\mathrm{T}}$. Note that there are no manipulated variable change-rate constraints here, which will degrade the control performance, particularly, the speed overshoot.

5.2.3   New MPC-Based Flux-Weakening Algorithm

In allusion to the problems shown in Sect. 5.2.2, the undermentioned measures are taken to establish the novel MPC-based flux-weakening algorithm to improve the classic one.

This part uses the first linearizing method in Sect. 5.2.2, where one of the most important priorities is ensuring that the load torque will be included in a LPM. Assume that a constant torque is imposed on the rotor shaft, and then the value of T_l is fixed. Regard T_l as a measured stable disturbance and add a constant term 1 into the state variable vector (5.20), after which the linear drive model can be described as (5.29) ~ (5.33). So far, the ILPM of IPMSM can be online modified after offline identification whenever the load torque varies during flux-weakening operation.

Looking at the model at length, an extra feature of (5.29) is that $i_d(k)i_q(k)$ is retained so as to be more appropriate for IPMSM with no need for the assumption $L_d = L_q$. As is illustrated in (5.30), the dq-axis currents are seen as the output variables. In this part, i_q is only a measured output that needs to be constrained and penalized during optimization process, but i_d as well as ω_m functions as the reference of MPC controller in marked contrast with the natural field-weakening algorithm. In detail, when a higher speed setpoint than basic speed ω_{bas} is given, a negative i_d requires to be controlled at a vested level, so that two tracking paths are wanted in the new algorithm.

$$\mathbf{x}(k) = [i_d(k), i_q(k), \omega_m(k), \overline{i_d i_q}(k), \overline{\omega_m i_d}(k), \overline{\omega_m i_q}(k), 1]^T \tag{5.29}$$

$$\mathbf{y}(k) = [i_d(k), i_q(k), \omega_m(k)]^T \tag{5.30}$$

$$\mathbf{u}(k) = [u_d(k), u_q(k)]^T \tag{5.31}$$

$$
A = \begin{bmatrix}
\frac{L_d - T R_s}{L_d} & 0 & 0 & 0 & 0 & \frac{T L_q P}{L_d} & 0 \\
0 & \frac{L_q - T R_s}{L_q} & -\frac{T \Psi_f P}{L_q} & 0 & -\frac{T L_d P}{L_q} & 0 & 0 \\
0 & \frac{1.5 T p \Psi_f}{J} & \frac{BT + J}{J} & \frac{1.5 T p (L_d - L_q)}{J} & 0 & 0 & -\frac{T T_l}{J} \\
0 & 0 & 0 & 1 & 0 & 0 & 0 \\
0 & 0 & 0 & 0 & 1 & 0 & 0 \\
0 & 0 & 0 & 0 & 0 & 1 & 0 \\
0 & 0 & 0 & 0 & 0 & 0 & 1
\end{bmatrix}
\tag{5.32}
$$

$$
B = \begin{bmatrix} \frac{T}{L_d} & 0 & 0 & 0 & 0 & 0 & 0 \\ 0 & \frac{T}{L_q} & 0 & 0 & 0 & 0 & 0 \end{bmatrix}^T, \quad
C = [] \begin{bmatrix} 1 & 0 & 0 & 0 & 0 & 0 & 0 \\ 0 & 1 & 0 & 0 & 0 & 0 & 0 \\ 0 & 0 & 1 & 0 & 0 & 0 & 0 \\ F_{4\times7} = 0 \end{bmatrix}, \quad D = 0
\tag{5.33}
$$

However, another issue that how to preset the reference value $i_d{}^*$ shows up. Considering that the IPMSM flux-weakening process is characterized by full voltage operation and the line-to-line back-electromotive force (EMF) approximately equals the DC-bus voltage when ignoring the turn-on voltage of IGBTs, the rotating speed ω_m and the air-gap flux linkage ψ'_f have the relationship expressed by

$$
\Psi'_f = \frac{U_{dc}}{\sqrt{3} C'_e \omega_m}
\tag{5.34}
$$

where C'_e is the voltage constant relevant to the motor and U_{dc} represents the DC-bus voltage. Meanwhile, the flux linkage in the motor is also described as

$$
\Psi'_f = \Psi_f + L_d i_d
\tag{5.35}
$$

Because the DC-bus voltage is usually unchangeable once the IPMSM system is established, the reference current $i_d{}^*$ is now merely related to the motor speed setpoint value $\omega_m{}^*$. More specifically, $i_d{}^*$ is supposed to be controlled by the following rules

$$
\begin{cases}
i_d^* = 0, & if\ \omega_m^* \le \omega_{bas} \\
i_d^* = -\frac{\Psi_f}{L_d} + \frac{U_{dc}}{\sqrt{3} C'_e \omega_m^* L_d}, & if\ \omega_m^* > \omega_{bas}
\end{cases}
\tag{5.36}
$$

Finally, the purpose of restraining the change rate of manipulated variables is to reduce the acceleration of shaft, lowering the dynamic speed overshoot, and meanwhile, that is conducive to preventing the motor and the IGBT from voltage and current surge to prolong the service life of the system, In addition to those commonly used constraints in (5.27) and (5.28), the new flux-weakening algorithm introduces one more constraint which is like

$$\begin{aligned} |\Delta u_q(k)| \le \Delta u_{q\,max} \\ |\Delta u_d(k)| \le \Delta u_{d\,max} \end{aligned} \qquad (5.37)$$

where $\Delta u_q(k)$ and $\Delta u_d(k)$ are manipulated voltage change rate and $\Delta u_{q\,max}$ and $\Delta u_{d\,max}$ are the maximal q-, d-axis control voltage variation during each time step, respectively.

The issue how to determine $\Delta u_{q\,max}$ and $\Delta u_{d\,max}$ arises. When the machine works under no load in the constant torque region where i_d and i_q are nearly zero, u_d in (1) roughly equals to zero as well. Then, u_q can be approximately regarded as the sole control voltage during start-up course, and it peaks (no larger than U_{dc}) when the IPMSM reaches the basic speed. Given that the required settling time of the system is t_s and u_q grows monotonously and uniformly to the maximum voltage U_{dc}, $\Delta u_{q\,max}$ is going to be approximated by

$$\Delta u_{q\,max} = \frac{T \cdot U_{dc}}{t_s} \qquad (5.38)$$

It is more complicated to decide $\Delta u_{d\,max}$ theoretically. Nonetheless, now that $\Delta u_{q\,max}$ in (5.38) can guarantee lower speed overshoot and avoid large voltage or current surge in the constant torque region, it is believed that it is the same with $\Delta u_{d\,max}$ in the constant power region, that is

$$\Delta u_{d\,max} = \Delta u_{q\,max} \qquad (5.39)$$

A block diagram of the proposed MPC algorithm realized on an IPMSM control system is shown in Fig. 5.8.

5.2.4 Verifications

5.2.4.1 Comparative Study on Control Performance

The improved MPC-based flux-weakening algorithm in comparison with a classic one is simulated to track forward command speeds in MATLAB/Simulink 2013b. The motor and control parameters of IPMSM are listed in Table 5.1. In this section, both no-load and under-constant-load conditions are analyzed.

Firstly, the machine speeds up from 0 rad/s to 90 rad/s, where i_d is supposed to remains zero and an external load of 4 Nm is applied at $t = 2.0$ s when the parameter T_l in the ILPM is modified to 4 simultaneously. Then, set the reference speed as 120 rad/s at $t = 4.0$ s, after which flux-weakening operation launches. Finally, the load is removed immediately when $t = 6.0$ s. The simulation results are shown in Figs. 5.9, 5.10 and 5.11.

(a) Analysis on steady-state performance

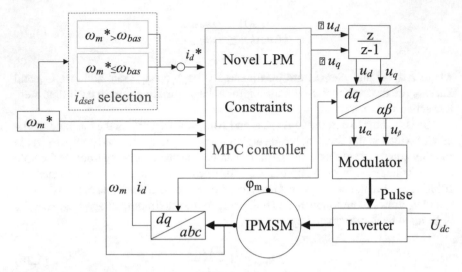

Fig. 5.8 Block diagram of the proposed MPC algorithm

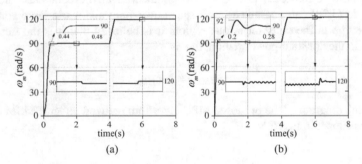

Fig. 5.9 Speed response. **a** Proposed MPC-based flux-weakening algorithm. **b** Classic flux-weakening MPC algorithm

The speed response results reflect marked differences between the proposed MPC-based flux-weakening algorithm and classic flux-weakening MPC algorithm although both schemes can track the commands well. Figure 5.9a illustrates the motor speed remains stable after it arrives at the setpoint no matter whether the IPMSM is loaded or not during the whole test range, but as is seen in Fig. 5.9b, the speed fluctuates slightly when the motor is under load status. This phenomenon shows that the ILPM can contribute to achieving higher accuracy, which can also be proved by the response of currents i_q and the torque characteristics. Another advantage of the new MPC-based flux-weakening algorithm is that the current i_d in Fig. 5.10a is controlled to stabilize at zero and -36.6 A when the speed setting point is 90 rad/s and 120 rad/s, respectively. Comparatively, the i_d in Fig. 5.10b is not as stable as that in Fig. 5.10a during steady-state operation, acting on the air-gap magnetic field and further clearly reacting on the control performance.

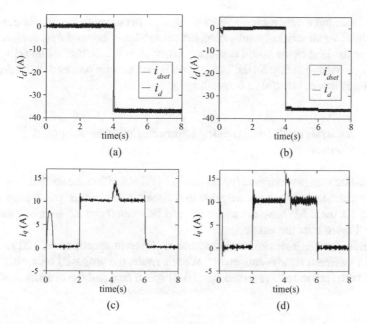

Fig. 5.10 Response of currents under two kinds of algorithms. **a** Proposed MPC-based flux-weakening algorithm. **b** Classic flux-weakening MPC algorithm

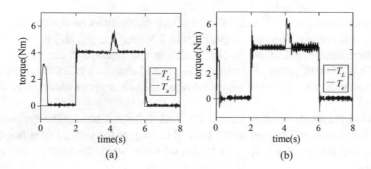

Fig. 5.11 IPMSM torque characteristics under two kinds of algorithms. **a** Proposed MPC-based flux-weakening algorithm. **b** Classic flux-weakening MPC algorithm

(b)   Analysis on dynamic performance

Dynamic performance evaluation of an IPMSM control system needs to consider three crucial aspects, namely, settling time, speed overshoot and current surge. In terms of start-up speed, the rise time of the proposed MPC algorithm is relatively longer (0.45 s) than that of the classic algorithm (0.25 s). Nevertheless, thanks to the reformative constraints in the proposed MPC-based flux-weakening algorithm, Fig. 5.9a shows lower speed overshoot than Fig. 5.9b. When focusing on the course of load step response, it is observed that the current i_q and the electromagnetic torque T_e

change faster but a noticeable damped oscillation occurs with the classic controller. Meanwhile, the simulation results demonstrate relatively larger d-axis current surges whenever the load or the speed command changes in Fig. 5.10b, while fairly stable current i_d is visible in Fig. 5.10a. That phenomenon exactly proves that the proposed algorithm is more powerful in controlling i_d.

5.2.4.2 Analysis on Speed Tracking Characteristics for Proposed Controller

The speed tacking properties of the proposed MPC algorithm are specially discussed in this section taking both acceleration and deceleration in flux-weakening region. In Fig. 5.12, the IPMSM works under no load between 0 and 2.5 s while a torque of 4 Nm is loaded over the left period.

Overall, the machine can remain stable at different speeds (e.g., 120 rad/s and 110 rad/s) without steady-state errors. What's more, the proposed controller is able to track both forward and reversed command speed fast, and the overshoot is hardly seen.

5.2.4.3 Impact of DC-Voltage Fluctuations on Stability and Accuracy of Proposed Controller

In order to discuss the influence of DC-voltage fluctuations on the control characteristics, a sinusoidal disturbance (amplitude 2 V, frequency 1 Hz) is overlaid onto the voltage source. As shown in Fig. 5.13, the machine speed can keep stable at 120 rad/s if DC-link voltage experiences an upward change, while a relatively higher fluctuation of 2 rad/s (1.7%) is witnessed when the voltage goes down. Honestly, the proposed.

MPC-based algorithm does not use DC-link voltage to predict the system states, so that the system will not get out of control as long as the voltage does not change sharply, and the speed can recover to the set point when the voltage fluctuations disappear.

5.2.4.4 Experimental Results

Apart from simulation, experiments are conducted on a three-phase IPMSM whose parameters are also consistent with Table 5.1. Firstly, steady-state performances of the improved algorithm and the conventional algorithm under no-load as well as load conditions are depicted in Figs. 5.14 and 5.15. It can be found that all of the speed waveforms nearly level off at 120 rad/s with no error, and i_d is about -37 a during flux-weakening operation. As a whole, higher fluctuations can be seen with load status, but Fig. 5.15a shows obviously lower speed and current volatility than Fig. 5.15b, which is consistent with the simulation results.

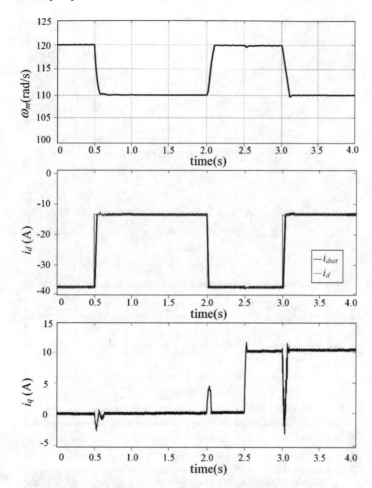

Fig. 5.12 Speed tacking properties of the proposed MPC controller

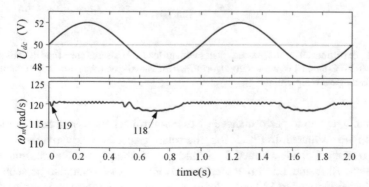

Fig. 5.13 Control performance of the proposed MPC controller considering DC-voltage fluctuations

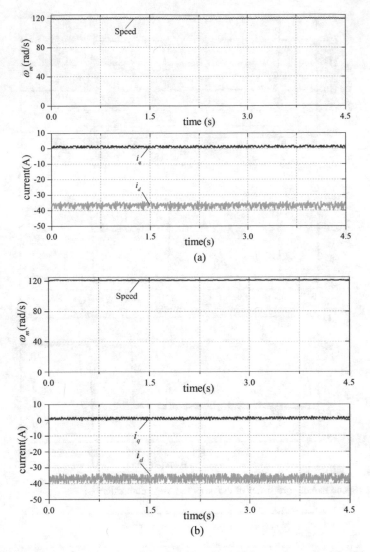

Fig. 5.14 Experimental results of steady states with no load. **a** Measured speed and dq-axis currents of proposed MPC-based flux-weakening algorithm. **b** Measured speed and dq-axis currents of classic flux-weakening MPC algorithm

Secondly, dynamic performances of two control schemes in speed-up mode of operation are evaluated. In Fig. 5.16b, the rotating speed quickly arrives at 90 rad/s with an overshoot of 3% within 0.3 s. i_q and T_e shift sharply and the maximum value of them are 10 A and 1.6 Nm at start-up. Overshoot (4%) can also be seen as the IPMSM is accelerated to 120 rad/s. In comparison, Fig. 5.16a shows the rise time of the proposed MPC algorithm is longer (0.45 s) but there exists no speed overshoot during speed-up process. During start-up, the maximum value of i_q and T_e are more

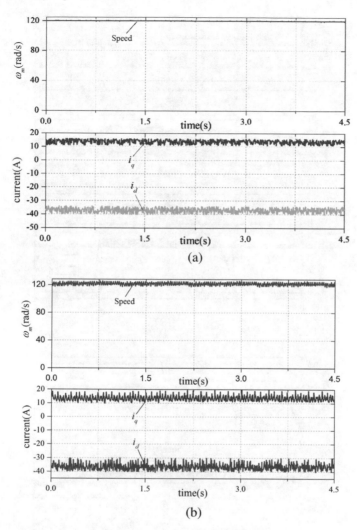

Fig. 5.15 Experimental results of steady states under 4 Nm. **a** Measured speed and dq-axis currents of proposed MPC-based flux-weakening algorithm. **b** Measured speed and dq-axis currents of classic flux-weakening MPC algorithm

stable and smaller than the classic MPC algorithm, 8.5 A and 1.2 Nm, respectively. One may notice that the d-axis current of classic flux-weakening MPC algorithm gets negative when a higher speed set point is given in constant torque zone while it is almost invariable for the proposed MPC-based flux-weakening algorithm, which is in agreement with the simulated results in Fig. 5.11a, b. Figure 5.16 proves that although both of the MPC algorithms are fundamentally for constant power zone, they are also well applied to constant torque zone (e.g. 90 rad/s).

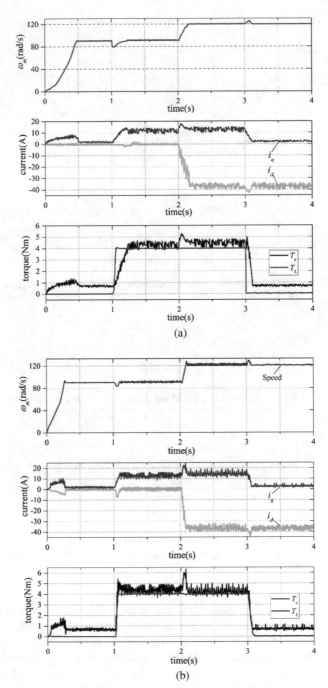

Fig. 5.16 Measured speed, *dq*-axis currents and torque waveforms response to start-up and load step-change before and after flux-weakening operation. **a** Proposed MPC-based algorithm. **b** Classic MPC algorithm

Then, the transient speed and current responses under abrupt change of load are investigated and presented in Fig. 5.16. As expected, when IPMSM is loaded with 4 Nm in constant torque zone, the speed drops visibly before it returns 90 rad/s, and conversely, ω_m might jump gently immediately when the load is moved in flux-weakening process, followed by getting back to 120 rad/s. What is indistinct in simulation but clear here includes that i_d suddenly becomes negative at both loading and unloading moments. However, the proposed algorithm presents smaller amplitude of variation (no more than 5 A).

Overall, the novel flux-weakening MPC algorithm presents slower current and speed tracking performances, but fixed settling time can be set, ensuring minimum overshoot. Another competitive advantage is that dq-axis currents keep more stable, especially at load condition.

The speed tacking properties of the proposed MPC algorithm are depicted in Fig. 5.17. Whether the machine operates under no-load or load conditions, the improved MPC algorithm is capable of tracking any command speed in the flux-weakening region efficiently, which is in accord with the simulation analysis in Fig. 5.12.

Finally, when DC-voltage has fluctuations, as is shown in Fig. 5.18, the control system can remain stable in terms of machine speed. After the voltage fluctuations vanish, the speed gets back to the set point quickly. What needs to be pointed out is that the experimental test witnesses higher speed fluctuation than simulation, with 0.8 rad/s when the voltage rises and 3.5 rad/s (2.9%) when the voltage decreases, respectively.

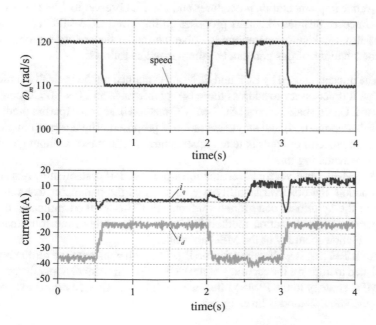

Fig. 5.17 Speed tacking properties of the proposed MPC-based algorithm

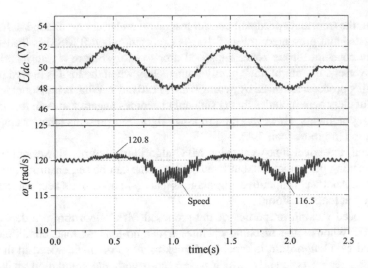

Fig. 5.18 Measured control performance of the proposed MPC controller considering DC-voltage fluctuations

5.3 Summary

This Chapter presents effective techniques to improve the prediction accuracy of the FCS-MPC methods used in IPMSMs. In detail, a new FCS-MPCC controller based on innovative implementation procedures and SM-FL observer to dilute the LCF and FLM influence. Besides, this part proposes an improved CCS-MPC controller for IPMSM flux-weakening operations to solve the model linearization problem. The main contributions of this part can be summarized as follows:

(1) The impacts caused by LCF and FLM are explained mathematically, indicating that it is necessary to address these two problems in an FCS-MPCC controller used. On this basis, a novel FCS-MPCC method based on tripartite predictions (sub-step method) and single actuation is proposed. The designed implementation procedures contribute to higher control performance without generating more switching loss.

(2) An SM-FL observer for FL estimation is presented. By establishing a Lyapunov function, the observer stability is analyzed and the parameter that can stabilize the system is determined theoretically. This method indicates that the parameter-identification method is able to improve both the robustness and prediction accuracy of the MPC controllers.

(3) An ILPM that is well-suited for the IPMSM with constant load torque is established to improve the accuracy of the CCS-MPC. In order to apply the proposed MPC strategy to the IPMSM flux-weakening operations, a new d-axis current generation method is investigated.

In this part, although the new MPC-based algorithms provide more precise IPMSM PPMs to achieve better performance characteristics, there still exist some limitations that should be addressed. Firstly, the method to calculate the d-axis current reference by use of the speed reference in Sect. 5.2.3 is not always efficient because it is generally parameter-dependent. So a MPC controller which incorporates brand-new i_d^* determination method without adopting extra PI controller should be investigated. Secondly, the predicting model in Sect. 5.2 only works effectively when the load torque is constant and known in advance, so that the proposed method requires prior knowledge of the load torque when controlling an IPMSM. Consequently, it is by far not widely applicable to all cases.

References

1. S. Driss, S. Farhangi, M. R. Nikzad, "Low switching frequency model predictive control of PMSM drives for traction applications," *2018 9th Annual Power Electronics, Drives Systems and Technologies Conference (PEDSTC)*, (Tehran, 2018), pp. 300–305
2. W. Hu, H. Nian, D. Sun, Zero-sequence current suppression strategy with reduced switching frequency for open-end winding PMSM drives with common DC bus. IEEE Trans. Industr. Electron. **66**(10), 7613–7623 (Oct. 2019)
3. X. Zhang, L. Zhang, Y. Zhang, Model predictive current control for PMSM drives with parameter robustness improvement. IEEE Trans. Power Electron. **34**(2), 1645–1657 (2019)
4. M. Yang, X. Lang, J. Long, D. Xu, Flux immunity robust predictive current control with incremental model and extended state observer for PMSM drive. IEEE Trans. Power Electron. **32**(12), 9267–9279 (2017)
5. Z. Song, F. Zhou, Observer-based predictive vector-resonant current control of permanent magnet synchronous machines. IEEE Trans. Power Electron. **34**(6), 5969–5980 (2019)
6. Z. Tian, C. Zhang, S. Zhang, Analytical calculation of magnetic field distribution and stator iron losses for surface-mounted permanent magnet synchronous machines. Energies **10**(320), 1–12 (2017)
7. A. Sabanovic, Variable structure systems with sliding modes in motion control—a survey. IEEE Trans. Industr. Inf. **7**(2), 212–223 (May 2011)
8. Y. Zhang, W. Cao, S. McLoone, J. Morrow, Design and flux-weakening control of an interior permanent magnet synchronous motor for electric vehicles. IEEE Trans. Appl. Supercond. **26**(7), 1–6 (2016)
9. X. Liu, H. Chen, J. Zhao, A. Belahcen, Research on the performances and parameters of interior PMSM used for electric vehicles. IEEE Trans. Industr. Electron. **63**(6), 3533–3545 (2016)
10. J.H. Park, D.J. Kim, K.B. Lee, Predictive control algorithm including conduction-mode detection for PFC converter. IEEE Trans. Industr. Electron. **63**(9), 5900–5911 (2016)
11. B.S. Riar, T. Geyer, U.K. Madawala, Model predictive direct current control of modular multilevel converters: modeling, analysis, and experimental evaluation. IEEE Trans. Power Electron. **30**(1), 431–439 (2015)
12. C.S. Lim, E. Levi, M. Jones, N.A. Rahim, W.P. Hew, A comparative study of synchronous current control schemes based on FCS-MPC and PI-PWM for a two-motor three-phase drive. IEEE Trans. Industr. Electron. **61**(8), 3867–3878 (2014)
13. M. Cheng, F. Yu, K.T. Chau, W. Hua, Dynamic performance evaluation of a nine-phase flux-switching permanent-magnet motor drive with model predictive control. IEEE Trans. Industr. Electron. **63**(7), 4539–4549 (2016)
14. T. Geyer, Model predictive direct current control: formulation of the stator current bounds and the concept of the switching horizon. IEEE Indus. Appl. Magazine **18**(2), 47–59 (2012)

15. Z. Mynar, L. Vesely, P. Vaclavek, PMSM model predictive control with field-weakening implementation. IEEE Trans. Industr. Electron. **63**(8), 5156–5166 (2016)
16. Y. Zhang, H. Yang, Two-Vector-based model predictive torque control without weighting factors for induction motor drives. IEEE Trans. Power Electron. **31**(2), 1381–1390 (2016)
17. A. Linder, R. Kennel, "Model predictive control for electrical drives," *2005 IEEE 36th Power Electronics Specialists Conference*, (Recife, 2005), pp. 1793–1799
18. L. Wang, *Model Predictive Control System Design and Implementation Using Matlab*, Advances in Industrial Control. (Springer, Verlag London, ch. 2, Mar 2009), pp. 47–48
19. S. Bolognani, L. Peretti, M. Zigliotto, Design and implementation of model predictive control for electrical motor drives. IEEE Trans. Industr. Electron. **56**(6), 1925–1936 (2009)
20. J. Guzinski, H. Abu-Rub, M. Diguet, Z. Krzeminski, A. Lewicki, Speed and load torque observer application in high-speed train electric drive. IEEE Trans. Industr. Electron. **57**(2), 565–574 (2010)
21. M. Preindl, S. Bolognani, Model predictive direct torque control with finite control set for PMSM drive systems, part 1: maximum torque per ampere operation. IEEE Trans. Industr. Inf. **9**(4), 1912–1921 (2013)

Chapter 6
Conclusion

Yaofei Han, Chao Gong⊙, and Jinqiu Gao

As an advanced control method, model predictive control (MPC) has been witnessing exponential growth in the application of AC motor control. However, it must be acknowledged that the technologies concerning MPC used in the AC motor drives are not entirely mature and there are still many problems that need to be studied deeply. Obviously, the robustness and accuracy issues of MPC are attracting significant attention from not only the engineers in the industry but also the scholars, making them valuable topics for research. In this book, the robustness- and accuracy-related issues of the MPC are addressed in detail. Especially, the parameter mismatch issues are solved by using both the disturbance compensation method and the parameter identification method. And the prediction accuracy of the MPC is improved by using new predicting plant models (PPM), delay compensation and linearization techniques. It deserves to be mentioned that, for the sake of comprehensiveness, this book not only focuses on one kind of Ac motor. Instead, the induction motors (IM), wound field synchronous motors (WFSM) and permanent magnet synchronous motors (PMSM) are the targeting research objects. The proposed techniques proposed in this research have been proven to be effective to improve the robustness and prediction accuracy of the MPC method. The main work and contributions of this book can be summarized as follows:

In Chap. 1, the development of the MPC theories is introduced firstly. It can be seen that the MPC method has been studied for nearly forty years, and it is a control method based on optimization. The knowledge and experiences concerning MPC are prompting it to be applied widely now. Secondly, the structures of the IMs, WFSMs and PMSMs are presented, illustrating the research objects in this book. Thirdly, the common problems of MPC used in the AC motor drives are detailed. These

Y. Han
National Maglev Transportation Engineering R&D Center, Tongji University, Shanghai 201804, China

C. Gong (✉) · J. Gao
School of Automation, Northwestern Polytechnical University, Xi'an 710072, China

© The Author(s), under exclusive license to Springer Nature Singapore Pte Ltd. 2022   127
Y. Han et al. (eds.), *Model Predictive Control for AC Motors*,
https://doi.org/10.1007/978-981-16-8066-3_6

demonstrate that there is still a long way to go before the MPC becomes mature. In the rest chapters of this book, the research mainly focuses on solving these problems so as to improve the robustness and accuracy of the MPC. Finally, the codes of one typical MPC method used in PMSM control are provided. Hope they can guide the new researchers to be familiar with the implementations of the MPC strategies as quickly as possible.

In Chap. 2, a disturbance observer-based strategy is proposed to reduce the impacts of the parameter mismatch issue on the control performance of IMs. Firstly, the chapter illustrates the implementation process of a finite control set model predictive current control (FCS-MPCC) algorithm used for the IMs. For the sake of high flux observation estimation accuracy, a new numerical-solution-based flux observer is introduced. In addition, the reasons why the performance of the traditional FCS-MPCC method is low considering the influences of parameter mismatch are explained. Secondly, a sliding mode disturbance observer is proposed, which is seldom studied in the field of MPC used for the AC motors. Simultaneously, the stability of the proposed sliding mode observer is discussed. Finally, the simulation results are shown to verify the proposed algorithms. It can be concluded that by using the proposed disturbance-rejection method, the robustness of the FCS-MPCC method used in the IM drives is improved.

In Chap. 3, a parameter-identification method is proposed to eliminate the impacts of the inductance mismatch problems on the control performance of the FCS-MPCC strategy used in the novel WFSMs based on capacitive coupling excitation. Firstly, this chapter presents the modeling method for the novel WFSMs, laying the ground for achieving the FCS-MPCC method and the inductance observers. Then, series of online parameter observers based on the SM principle are developed, with their robustness and stability discussed. This illustrates the proposed observers and observer-based FCS-MPCC method is more suitable for the low-d-axis current situations. In other words, more work is needed to extend the applications of the proposed strategies. Finally, the simulation results of the proposed algorithms are given to prove that they are well-suited for the low-d-axis current situations. Undoubtedly, this chapter introduces another effective solution to improving the robustness of the FCS-MPCC method used in the AC motor drives.

In Chap. 4, the accuracy of the MPC methods are focused on. Especially, two problems influencing the prediction accuracy of the FCS-MPCC method are solved. That is, the traditional PPM used for prediction in the low-control-frequency situations is not accurate and it is difficult to accurately achieve a multi-objective strategy with speed and currents combined. Firstly, this chapter firstly proposes a numerical-solution-based PPM, which has higher prediction accuracy than the traditional. Secondly, a multi-objective finite control set model predictive control (FCS-MPC) method (combing speed and current characteristics) is developed. The calculation delay and weighting factor tuning problems are solved to improve the accuracy of the MPC. Finally, both simulation and experimental results are given to validate the proposed methods. It deserves to be mentioned that, the proposed strategies not only provide guidelines to improve the accuracy of the MPC methods but also enrich the theories of MPC used in motor control.

In Chapter 5, the problems that the traditional PPM in the low-control-frequency situations is not accurate and the flux linkage mismatch issue is inevitable during control are focused on, which are in different perspectives from the other chapters. In addition, the nonlinear model of the PMSMs is elaborated to make it more suitable for MPC control. Firstly, this chapter explains the reasons why the performance of the traditional PPM is inaccurate more clearly than that in Chap. 4. Then, a flux-observer-based sub-step FCS-MPC method is developed. Thirdly, an MPC method with a linearization model is proposed for flux-weakening control. The simulation and experimental results are compared finally to prove that the proposed technologies are effective to improve the accuracy of the MPC controllers.

All in all, in terms of the topic of robust and accurate MPC methods, this research combines both theoretical analysis and simulation/experimental validation to investigate and develop comprehensive MPC theories and techniques for the AC motors. In this area, the following achievements have been made in this study:

- Disturbance observation and compensation methods and parameter identification methods are developed to improve the robustness of MPC methods against electrical parameter variations of the AC motors.
- Numerical-solution-based and sub-step PPMs are developed to enhance the prediction accuracy of the FCS-MPC methods.
- Novel flux linkage observation, delay compensation and linearization techniques are proposed to improve the accuracy of the MPC methods.

Printed in the United States
by Baker & Taylor Publisher Services